Beachcomber's
Guide to FLORIDA
MARINE LIFE

Beachcomber's Guide to FLORIDA MARINE LIFE

William S. Alevizon

Gulf Publishing Company
Houston, Texas

To all beach bums—past, present, and future

Gulf Publishing Company
Book Division
P.O. Box 2608 • Houston, Texas 77252-2608

10 9 8 7 6 5 4 3 2 1

Printed in the United States of America

Library of Congress Cataloging-in-Publication Data

Alevizon, William S.
  Beachcomber's guide to Florida marine life/William S. Alevizon.
       p. cm.
    Includes bibliographical references (p.) and index
    ISBN 0-88415-128-X
    1. Seashore biology—Florida. 2. Marine biology—Florida. 3. Beaches—Florida. 4. Beachcombing—Florida. I Title.
QH105.F6A49   1994
574.9759′0914′6—dc20   94-17568
                                                          CIP

Contents

Chapter 9
The Keys: Florida's Tropical Isles 158

Acknowledgments

The author wishes to thank Joann Mossa and Walter S. Judd, colleagues at the University of Florida, for their thorough chapter reviews and many useful suggestions. Deidre Spayd and Trish Petty were instrumental in bringing my illustration concepts to life.

Figure Credits

Illustrations

Susan Brunenmeister, Ph.D.: Figures 3-16, 3-18
Nick Fotheringham, Ph.D.: Figures 3-7, 3-10, 3-11, 3-12, 3-13, 3-14, 3-19, 5-4, 5-7, 5-10, 5-12, 5-13, 5-14, 5-15, 5-16, 5-17, 7-1, 7-2, 7-3, 7-4, 7-8, 9-3, 9-4
Patsy Menefee: Figures 3-8, 3-9, 3-15, 3-17, 5-3, 5-6, 5-8, 5-9, 5-11, 6-3, 6-4, 6-5, 6-6, 7-5
Trish Petty: Figures 2-2, 2-9, 2-11, 2-12, 2-13, 3-1
Ruth Rasche: Figures 9-1, 9-2
Diedre Spayd: Figures 1-5, 2-10, 6-2, 6-7, 6-8, 7-7, 7-9, 7-11

Black/white Photos

Florida Power and Light: Figures 8-1, 8-2, 8-4, 8-5

Color Plates

Holly Hart: Plates 3, 4, 6, 7, 8, 9, 10, 25, 40, 41, 42
Paul Hart: Plates 11, 27, 43
Walter S. Judd: Plates 12, 13, 16, 17, 18, 19, 20, 21, 23, 24
Mike Marshall: Plate 45
Diedre Spayd: Plates 29A–F, 31A–E, 33A–E, 36A–C

Preface

At the place where land and sea merge exists a unique and special realm. For most of our planet's creatures, including ourselves until just a comparatively short time ago, this thin bright line we call the beach marks an inviolate boundary between a familiar home and a tantalizingly close but forbidden world; very few creatures are equipped to visit both land and sea.

Like no other place so accessible to so many of us, beaches open the door to our imaginations and our most private dreams. Here, our fantasies of unknown worlds and untasted adventures are nourished by shadowy shapes gliding quickly from view into unseen depths, or by haunting clues teasingly tossed ashore by the sea herself. Perhaps it is this sense of mystery, as well as the respite offered from a world too crowded with the scars of our passing, that leads to our obsession with beaches. We are drawn again and again, but do not know why.

Few places in the world match Florida when it comes to beaches. The Sunshine State boasts a coastline of 1,800 miles, most of which is lined by beautiful sandy shores caressed by warm winds and seas, and harboring a fascinating assortment of beach life. This book explores the world of Florida's beaches, and the creatures that dwell there. It is intended to enhance the experience of visiting and for a time losing oneself in them.

Bill Alevizon, Ph.D.

1

The Basics of Beachcombing

BEACHCOMBING AS A STATE OF MIND

Beachcombing may be described as the human pastime of exploring the edge of the sea. No one knows how or when beachcombing began, but it was surely a very, very long time ago—well beyond the range of recorded human events. If recent speculations about our distant past are essentially correct, then the first beachcombers-to-be must have inadvertently stumbled upon the Indian Ocean a few million years ago.

It is tempting to wonder what those first brave, naked explorers must have felt upon seeing the endless waters stretching off beyond the horizon into infinity, and hearing the roar and power of the crashing surf. Utter amazement? Awe? Fear? Probably all of the above. But, as with their brethren of today, they also must have felt an irresistible urge to explore this new world, with its promise of unknown treasures. It must have been more than a bit scary to be the first to wander beyond the concealing dunes and out onto the open beach. On the one hand, nothing very big could sneak up on you out there. But on the other, if something dangerous *did* catch you unaware, there was no place to hide, nothing to climb—there would be only one escape. Perhaps it was in this very way that swimming in the sea also first began! At any rate, exploring beaches must have soon become a regular part of human existence, and has remained so to the present.

There is without question much more to beachcombing than merely scouring the landscape for something unexpected, unknown, or possibly useful. While the simple pleasure of finding a shell to bring home may at first be the main goal, the novelty of collecting trinkets soon wears off. Before long, finding anything noteworthy becomes more or less irrelevant—it's just *being there* that counts.

So, beachcombing is much more than a souvenir hunt. To *aficionados,* it is more a state of mind than an activity. In part, it involves observing seashore life and its traces. In part, it involves watching and listening to the sea in all her moods. In part, it involves feeling the wind and the sun and the sand. But in total, beachcombing transcends the simple sum of these individual facets. Somehow, their unique combined effect upon our bodies, minds, and spirits allows our usual preoccupation with matters of absolutely no consequence to any of Earth's other creatures to quietly drop away. It is then, for a brief time, that we become the quintessential beachcomber, just another animal in harmony with its environment (Figure 1-1).

1-1. Beachcombing brings out a hidden spirit in all of us.

The special magic hidden in beaches that allows us to open such a door has inspired generation after generation of devotees, and created kindred spirits among such seemingly diverse factions as beach-strollers of 100 years ago and surfers of today. In this sense, beachcombing is to all intents and purposes a form of personal meditation that might even be called an art. But unlike most other forms of meditation, beachcombing requires no special knowledge or training; the beach itself creates the proper mood. From this perspective, one might, in a very real sense, participate in the essence of beachcombing while strolling, sunbathing, fishing, or even surfing.

Although one needs nothing but an open mind to become a happy beachcomber, each who ventures to the sea's edge, whether first-time visitor or dedicated beach-bum, invariably ponders things seen but not recognized or understood. What is that strange creature? Why is the beach so much wider today than it was two months ago? Where do the waves come from? These and many other questions are a regular part of beachgoers' thoughts. To those that make the effort to learn some of the answers, the beach experience is enriched. One never need equate learning about the beach with losing some of the mystery found therein, for each answer invariably brings with it even more questions.

BASIC BEACH OUTFITTING

As with any sport or outdoor activity, the experience of beachcombing may be made considerably more enjoyable and successful through proper preparation. There are several items that anyone planning to spend more than a few minutes on a Florida beach will be advised to have readily available. Although it may seem a bit of a nuisance to spend a few minutes gathering and packing your beach gear, you will soon be quite satisfied that the time was well spent. Hard-core beach bums usually keep a full stock of such basic items in their cars, so they are always ready to hit the beach.

First, let's consider clothing—what to wear to the beach. There are more aspects to this than might meet the eye, for proper beachwear is a blend of comfort, functionality and, for most people, style. This last factor—style—is the most intangible of the three. Some people put a high premium on fashion, while others are content with just about anything that doesn't cause outright embarrassment. Most beachcombers will fall somewhere in between these extremes, and for them there exists a wide range of possibilities when it comes to beachwear. Here we will suggest some elemental considerations in selecting your beach wardrobe.

The swimsuit is undoubtedly the pivotal item of beach attire. Modern swimsuits come in an amazing assortment of fabrics, colors and styles, and are available over a wide price range. It is suggested that you plan to buy your swimsuit(s) at your Florida beach destination rather than in your home town in another state. This suggestion is not made to enhance Florida revenues, but rather because the selection and even the prices are likely to be better where so many swimsuits are sold. Another advantage of this strategy is that before you take your first stroll along the sands, you will have a chance to see for yourself what is hot and what is not in this year's Florida beach scene, and thereby avoid the problem of feeling out of fashion, if you are sensitive to that sort of thing.

The main non-style considerations in choosing a swimsuit are straightforward. Above all, you will want something that is comfortable—not too tight or too loose.

Another major comfort consideration is drying time. It is better to select lightweight, quick-drying synthetic fabrics like nylon than the heavier cottons and the like. Wet suits soon irritate the skin, and many fabrics that might look great on a display rack are not really practical if you intend to hit the water. Color is largely a matter of personal preference, but keep in mind that dark colors absorb much more of the sun's heat than do light colors. This is not so much as consideration during the cooler winter months, but the summer sun in Florida is of tropical strength.

Along with your swimsuit you will want some sort of top to cover your shoulders and torso. This is a useful item in any season, as you may need additional warmth or sun protection. In either case, a top will probably allow you to comfortably remain at the beach longer than otherwise. Maximal protective benefit from the sun's rays and heat is gained from lightly textured, loose fitting white garments, particularly light cottons.

Eye protection is absolutely essential. On tropical beaches, the combination of the sun's direct rays along with sunlight reflected off both the sand and water causes a glare unmatched in almost any other environment. Sunglasses, as with other beach wear, come in a truly mind-boggling variety of styles and prices. Low end glasses begin at around $6, and usually offer reasonable protection against some of the most damaging ultraviolet radiation. However, the optics, fit, and durability of cheaper sunglasses are frequently poor, and vision may be distorted. Cheaper glasses also tend to scratch easily, a serious consideration considering the intended frequent exposure to highly abrasive quartz sand particles.

While it is possible to pay considerably higher to accommodate a craving for style, top-quality, functional glasses will cost anywhere from about $50 to $100, depending upon the features and materials. In most cases these are comfortable, sturdy, and offer a high degree of protection from virtually the entire harmful portion of the solar spectrum. A most useful feature for viewing the sea surface and below is polarization, a feature that provides the best contrast of variations in water colors.

Regardless of what you pay for sunglasses, get a pair with polarization. You will have better eye protection and durability, and the best possible view of all the things you came all the way to Florida to see. It is also wise to bring a hat of some sort. A simple baseball cap is ideal, but there are a wide variety of possibilities. The real essentials are that the hat will stay on your head in a stiff sea breeze, and has a sufficiently wide brim to shield your eyes, because sunglasses alone will not provide full shading.

The final item in your beach wardrobe is footwear. Sand absorbs a great deal of the sun's heat, and walking along the summer or spring beach can be a painful experience to bare feet. Here, the best choice by far is sandals of some sort. Select something waterproof and relatively inexpensive, so you needn't worry unduly about the possibility of damage or loss. Simple rubber or synthetic sandals known in beach circles as "thongs" or "slaps" are ideal.

It is no coincidence that the first choice for beachwear of dedicated beach people has changed relatively little over the last 30 years (Figure 1-2), for it incorporates all of the sensible principles outlined above. A lightweight swimsuit, light-colored cotton t-shirt, ball cap or visor, good pair of "shades" and a well worn pair of "slaps" will probably continue to remain the mainstays of knowledgeable beach people throughout the world for many, many years to come.

1-2. Classic beachwear never goes out of style. Useful accessories include a full size towel, backpack, and drinking bottle.

Accessories are also a necessary part of the successful beach experience. Probably the first and most essential of these is a good sunblock. Choose a "waterproof" type of appropriate protective strength for your particular needs. Extra protection may be useful for lips and nose if you are prone to chapping or burning in these most exposed areas.

An extra word of caution regarding sunburn is in order here, because sunburn is by far the most common injury received by Florida beachcombers. Enough has been said during the last few years about the possible harmful effects of too much ultraviolet radiation, and we need not belabor that point here. What should be stressed is that you need to use that sunblock you so carefully picked out, and use it often. Even the so-called "waterproof" sunblocks become seriously diluted or removed after swimming, and with a cool sea breeze and frequent wet-downs in the sea, it is quite likely that you may not be aware that you may be receiving a damaging dose of sunlight. The skin has a "delayed reaction" to the sun's harmful rays, and the full effect of sunburn will not become apparent for hours after you leave the beach, usually not until late that evening. Therefore, just because you look at your arm or look in a mirror while at the beach, and do not see any unusual change in skin color, it is *not* safe to assume that you have not yet been affected or that you have not already had too much sun for one day.

There is no simple way to judge this on any given day, because the degree of sunburn incurred depends upon a number of interactive factors, including most notably the amount of sunlight received, a person's particular sensitivity to sunlight, and the degree of protection afforded by prior tanning. The only sensible rule-of-thumb is to start out slowly with short visits to the beach and heavy sunblock, until you are able to safely extend your visits by becoming familiar with your own personal tolerance for sun exposure. If you stay at the beach until you noticeably turn pink, you will be in for a painful and possibly dangerous lesson.

A related concern, particularly during summer, is the possibility of serious dehydration. First-time visitors to the tropics seldom have any idea of how much water the human body needs to keep itself from overheating on a hot summer day, particularly when temperature effects are enhanced by exposure to the sun and wind, and immersion in the sea. It is wise to drink plenty of fluids before leaving home, and to replenish those regularly while at the beach—whether you feel thirsty or not.

A full-size beach towel is also a good investment. It will keep you relatively sand-free and tend to keep you from losing things like car keys and the like, which seem to always be attempting to escape by burying themselves in sand.

One final basic item worth considering is something in which to conveniently bag all your beach gear and personal items. A simple backpack of the type used by students throughout the world is ideal for this purpose. It is large enough to carry virtually everything mentioned above, including beach towel and sandals. It also has separate pockets in which critical items like car keys and wallet may be kept securely tucked away, rather than simply jumbled in with your towel and t-shirt (a mistake that often leads to keys, rings, and loose change being inadvertently dumped and lost in the sand as larger items are being removed). Another major advantage of the backpack is that it allows you to easily take all your belongings with you as you beachcomb, and still leave both hands free to examine things of interest.

BEACH SAFETY

Florida's beaches are not inherently dangerous places, but there are a few precautions one should consider to reduce the likelihood of an unpleasant experience. Common sense tells us that if we are to venture from the safety of our living rooms and into the natural world, we must take much of the responsibility upon ourselves to learn how to take care of ourselves in such situations. It is probably always safer to swim on a well-populated beach with lifeguards on duty, but only a tiny fraction of Florida's coastline is so protected at any given time. Even when lifeguards are present, they are not able to keep a close and constant watch on every person in the water. It is a big mistake to assume that just because there are lifeguards and other people around, it is alright to relax your guard. Rather, it is best to consider others around you as a measure of extra protection added to your primary insurance— your own knowledge, abilities, common sense and adherence to sound beach safety practices.

Perhaps the foremost beach safety tip is to know your own capabilities in the water, and never underestimate the power of the ocean. The southeastern coast of the United States is generally known for it comparatively tranquil inshore waters (Figure 1-3),

1-3. Florida is famous for the generally tranquil seas bordering the entire state.

and waves of such size that they would be considered large and dangerous on a world-wide scale are extremely rare here. Still, there is a huge amount of energy contained in even what might appear to be a "small" breaker, and if you are going to play in the waves of the surf zone, use extra caution until you gradually become adjusted to their effects and learn how to best deal with them.

Experienced beachgoers know that the first thing to do when arriving at the beach is to simply take a few minutes to carefully observe the sea before leaping headlong into it. See what the ocean is doing today. How big are the largest waves, and what is the duration in between "sets"? Wave trains (see Chapter 2) consist of a variety of wave sizes and shapes, and there is often a five-to-ten-minute interval between large sets of waves striking otherwise calmer shores. In the same vein, it is well to keep alert and scan the water around you while in the sea. Most of your time should be spent facing seaward—that's where waves and boats will be coming from. Be alert for the presence of longshore and rip currents as well. These may often be discerned before entering the water by looking at the speed and direction of drifting seaweed or debris, and by the presence of unusual patterns of surface turbulence. (See Chapter 2.)

What is the sea carrying today? Sight of the unmistakable purple-blue float of the Portuguese man-of-war in the water or along the beach will warn you to be especially alert while swimming that day. At times, particularly after large storms have battered nearby coastlines, breaking waves may contain sizeable pieces of driftwood or man-made debris (Figure 1-4), objects that even a small breaker may propel with sufficient force to cause serious bodily harm.

1-4. Large pieces of driftwood, often floating almost entirely submerged, may be tossed about the surf zone with great force until being cast up on the beach.

Are schools of fish frequently breaking the surface in an apparently agitated state? This may be a good indication that even larger predators are on the hunt nearby, which brings us to the subject of sharks. Of all the things a visitor to Florida's beaches might be concerned about from a safety standpoint, being bitten by a shark is among the least worthy. Unless you do something really stupid, you probably have a much better chance of winning the grand prize in the Florida lottery than being mauled by a shark during your Florida vacation.

Sharks are not infrequent visitors to Florida waters, but they are not the vicious troublemakers that some would have us believe. By and large, sharks spend their time simply going about their own instinctive business, which does *not* include hunting down and devouring swimmers. The only time humans are apt to be bitten is when they enter an area in which a shark is already feeding, and when the water is murky, causing the shark to mistake a part of a person for its most common natural prey—other fish. Even under these circumstances, the probability of a negative encounter with a shark is very rare.

Reasonable precautions include avoiding swimming amidst agitated schools of fish, or right in front of concentrations of surf fishermen. Leaving the water if you are bleeding or if you spot the dorsal fin of a shark nearby are also good ideas. There is a good chance that the owner of the dorsal fin you or other swimmers spotted is not a shark at all, but rather a close relative of Flipper. Dolphins are quite common along Florida's beaches and might well be mistaken for sharks by the inexperienced. The physical characteristics and behavior of these friendly mammals is described in Chapter 7, and allows them to be distinguished from sharks even at a distance.

Lightning may at times pose a serious threat to Florida beachcombers, and should always be taken seriously. Florida has more lightning than any other state, and victims are struck each year, sometimes at the beach. The highest frequency of thunderstorms is during the hot summer months, but the possibility exists virtually throughout the entire year. Because lightning and its accompanying thunder may be easily seen and heard for miles, there is seldom any excuse to be caught out in the open in the middle of a Florida thunderstorm. To avoid that unpleasant experience, just be aware of the changing sky around you, and take shelter early when a thunderstorm approaches (Figure 1-5). Thunderstorms

1-5. Approaching thunderstorms may be seen and heard for miles. Take shelter early.

frequently move quickly across an area, and if you wait until the rain begins to fall before you seek cover, you have waited too long. Extra vigilance is required if you have travelled far down the beach and have a two- or three-mile walk between yourself and the nearest shelter. If you should be caught too far from shelter, the recommended defense is to lie flat on the ground in a depression of some kind if available, and avoid proximity to trees or other tall objects. Stay out of the water, because sea water is an excellent conductor of electricity.

A final note of caution and common sense has to do with where to leave your car and valuables while you beachcomb. Most of Florida's more than 1,000 miles of beaches are not adjoined by streets or parking lots, and on occasion you may wish to explore some distance from a crowded beach. It is advisable to take a few precautions if that is your intent. One of the most obvious is to stop at a nearby store or other establishment and ask where you might safely leave your car while you walk. Another easily indulged precautionary measure is to call a local office of law enforcement—police or sheriff—and verify the advisability of your intent. In general, it is unwise to simply pull off along a deserted stretch of beach highway, lock your car, and walk on out of sight. Unfortunately, there are persons who make a living by preying on such trusting victims. With a small amount of effort, it is usually possible to find a safe place to park within a few miles of just about any beach you will be likely to explore. Serious beachcombers consider a walk of that distance as just about what they had in mind anyway.

COASTAL CONSERVATION GUIDELINES

We all share a responsibility to see after the welfare of the natural world around us, and all of its living things. Those who love the beach and the sea are inherently protective of these treasures, and seldom need to be motivated to practice and support sound conservation practices. In this section, we will focus on some effective ways in which individuals may assist in conservation efforts aimed at preserving the beauty and ecological integrity of the special world found only at the edge of the sea.

Positive conservation action may be mediated by anyone through two broad possible avenues—advocacy and personal example. Advocacy refers to involving oneself in conservation at the social and political levels. There are ample opportunities to take an active role in this vital process, either through direct support of conservation groups working towards these ends or through solicitation of elected and appointed government officials charged with administering relevant programs. Coastal zone management and natural resource protection are active areas in America today, and with relatively little effort, such as regularly reading the newspaper and watching local and national news broadcasts, one may be kept reasonably well informed of upcoming decisions that are likely to have major impacts upon the coastal zone and its life. Informed voting is one of the avenues available to effect positive conservation with minimal effort.

If you choose to provide financial support to a private conservation group, spend a bit of time to carefully review the credentials of your possible choices to make sure your money is used most productively for the cause intended. Be objective, and don't rely on slick promotional brochures or literature alone to decide which group or groups to support. During the last decade, the number of conservation groups soliciting support has increased dramatically, and some are unquestionably much more effective than others in achieving their stated goals. What types of activities does the group conduct? What specific achievements are they able to legitimately claim major credit for? What part of the annual budget goes to salaries, administrative costs (overhead) and fund raising, as opposed to actually conducting conservation activities? What facilities does the group possess, what is the size of the staff and what are their qualifications? You have every right to gather and evaluate such information before you send in your check. Top-notch conservation groups will gladly provide this to you, as they are quite proud of their track records and capabilities. Be suspicious of those who appear hesitant to provide straightforward answers.

At the level of personal example, there are a number of things that beachcombers should be aware of and conscientiously practice in the spirit of conservation. First, make an effort to learn and respect existing regulations regarding things you may legally do and not do when visiting Florida's beaches. Public beaches will generally have signs posted to inform visitors of such regulations (Figure 1-6). If you have any questions, call the nearest office of the Florida Marine Patrol for an up-to-date answer. This is particularly necessary if you intend to collect anything and take it with you. Many types of seashore and marine life are protected by state and/or federal law, and these laws are often backed by heavy penalties. Remember, ignorance of the law is not an acceptable legal defense, so if in doubt, ask first.

1-6. Read and observe posted conservation regulations.

The same conservation rules you see posted at public beaches should be equally applied to unposted beaches. Remember that you are just one of millions that will be visiting Florida's beaches that year, and a slight amount of damage multiplied by those kinds of numbers adds up to a major impact. Stay off the dunes and refrain from picking dune vegetation (Figure 1-7). Walkways that permit you to comfortably cross the dunes without causing environmental damage have been constructed at strategic locations throughout the state.

1-7. Destabilization of the dunes may result from visitors trying to gain beach access the easy way.

It is a matter of simple consideration as well as good conservation to make an effort to leave the beach as you found it (Figure 1-8). If you leave your trash behind on the sand, it not only makes an unsightly mess, it also becomes a danger to shore life and sea life. Plastic is easily carried out to sea by the next high tide, and

1-8. Dispose of your trash properly—it may be a hazard to wildlife.

will remain intact for centuries in the sea where it may entangle or be eaten by marine creatures with lethal effects.

Learn to respect all life-forms, no matter how humble they might appear. The best rule-of-thumb regarding wildlife is to look from a respectful distance while interfering with their behavior as little as possible. If you pick up something alive to examine it, first note exactly where it was and how it was positioned, and try to return it the same way. What might seem to you a relatively minor difference in position or orientation may mean life or death to simple seashore creatures.

It is not a good idea to attempt to feed wildlife either, no matter how benign your intentions. Wild animals all have certain natural foods that they are best adapted to use, and our prepared processed snack foods may not really be healthy nutrition for them (or us). Wildlife will often gladly take advantage of unnatural feeding opportunities, but these may lead to lasting behavioral changes (Figure 1-9) that may prove counterproductive or dangerous to the animals in the long term.

1-9. These seagulls have adopted a feeding pattern that includes cleaning a supermarket parking lot located some 60 miles from the coast.

If you go to the beach to watch sea turtles nesting, keep your distance as the turtle tests the location and decides whether or not to nest. The presence and activities of observers may in themselves deter nesting. Do not shine lights or pop strobes in the face of the turtle, blinding it temporarily. Once the turtle begins to dig the nest, she may be approached to within a few yards without interference. But refrain from touching the animal and shining lights in its face at all times.

Good conservation begins with each one of us. By knowing and observing good conservation practices ourselves, encouraging others to observe them, and by reporting observed violations, the average person may make a valuable contribution to the long-term health of the beaches and sea life of Florida.

2

Florida's Coastal Environment and Habitats

The edge of the sea is a dynamic region—an ever-changing physical environment inhabited by a diversity of life forms. This chapter provides an overview of the general structure and organization of the coastal environment, and some of the most influential forces that shape it.

COASTAL HABITATS: ZONATION AT THE EDGE OF THE SEA

If you observe the coastline from an elevated vantage point, you are immediately struck by a number of adjoining areas with a distinctly different appearances (Figure 2-1). Here, the bright green of dune plants, there the blinding white glare of beach sand. Beyond lies a foaming line of breakers, and beyond that the deep blue sea stretches unbroken to the horizon. Clearly, these must be very different kinds of places for creatures to live, and therefore must be inhabited by very different assemblages of living things.

While this inference holds true to some extent, it is also somewhat a misleading artifact of our own perception (or mispercep-

2-1. Looking seaward from the dune, the pattern of coastal zonation is clearly visible.

tion) of things. Many animals regularly use, for similar or different purposes, a number of what we might define as distinctly different kinds of living places (i.e., "habitats"). If we looked closely at the base of the dunes for example, we would see numerous animal tracks leading between the dense vegetation and the sand of the open beach. If we followed the daily movements of inshore fishes, we would also see that a number of species move freely and regularly between the waters below the breaking waves and the deeper quieter waters beyond. To many of the area's residents, such seemingly distinct places are not different worlds at all—just different rooms in the same house. They do not see their world from the vantage point of a map or an aircraft window—they live in it.

Thus, for practical purposes, this book follows the traditional approach of defining and discussing coastal "habitats" one at a time, but keep in mind that our divisions of the natural world are primarily for our own convenience, and often mean as little to marine creatures as do international boundaries. The biosphere is really more a *continuum* of interactive living and non-living parts than a *mosaic* of discrete organisms, habitats and ecosystems. The fact that we humans need to break nature up into bits and pieces to try and understand her does not mean that she actually exists in that way.

With these considerations in mind then, we will now take a look at the structure of the coastal environment as scientists traditionally view them. In this scheme, the beach is generally considered but one of a series of four ecologically linked *zones* that occur as parallel bands along coastlines. A diagramattic representation of the overall zonation pattern and the main structural features of each zone are illustrated in Figure 2-2. We will now examine their composition and major characteristics.

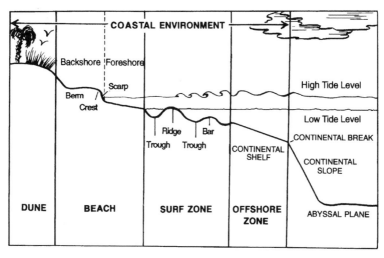

2-2. Diagram of coastal zonation.

The Offshore Zone

Starting to seaward, the coastal environment begins with a region called the *offshore zone,* consisting of a broad, relatively flat submerged platform called the continental shelf, and the waters that overly it. This is the only one of the four zones described that is not accessible to the wading or strolling beachcomber, although some of its creatures frequently are cast ashore by the sea. The seafloor of the continental shelf descends quite gradually over a substantial distance, eventually terminating at the *shelf break*—the point at which the seafloor begins an abrupt plunge into the abyss. Off much of the Atlantic coast of the United States, this occurs about 100 to 200 km offshore, and at an average depth of about 200m. There is considerable variation in this generalized pattern however. For example, in southern Florida, particularly off the Florida Keys, the offshore zone is much narrower. Here, the seafloor drops quickly into the depths of the Straits of Florida at distances of only 20–25 km from shore.

The Surf Zone

Between the inner margin of the offshore zone and the beach lies a turbulent area where most waves break (Figure 2-3). The outer edge of this line of breakers marks the beginning of the next

2-3. The surf zone—last stop for ocean waves.

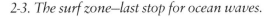

major coastal habitat, called the *inshore* or more descriptively, the *surf* zone. This is a particularly dynamic region that often includes a series of longshore (parallel to the shore) elevated areas of the seafloor called *bars* and *ridges,* separated by longshore depressions in the seafloor called *troughs.* The difference between a bar and a ridge is that the latter is exposed at low tide, whereas the former remains submerged. Bars, ridges, and troughs are constantly being reshaped by the pounding waves and changing tides, and by currents created in part by the presence of these features themselves. The surf zone is entirely submerged, except for its ridges and inner margin during unusually low tides.

The Beach

The next major coastal zone is that of the *beach* proper, sometimes also called the *shore.* Sandy beaches are generally subdivided into two broad regions: the *foreshore* and *backshore* (Figure 2-4). The foreshore is that area of the beach that lies between

2-4. The dry flat berm of the backshore (right) contrasts clearly with the wet sloping sands of the foreshore (left).

the daily high and low tide marks. The foreshore is an extremely transient habitat in a constant process of being alternately submerged and exposed by waves and shifting tides. It is the part of the beach primarily shaped by water movement.

The backshore, in contrast, is by definition beyond the reach of the daily tidal fluctuation. This area lies between the high water mark and the seaward edge of the dunes, and is shaped by the wind rather than the sea. Over time, the layers of sand that build the backshore tend to become rather evenly distributed, creating a characteristic broad flattened terrace, or *berm,* that is relatively level or slopes slightly towards the land. The berm becomes particularly well developed during summer, when gentle winds and waves result in a net transport of sand from longshore bars onto the beach. During winter, however, this process is reversed. Frequent storms with large waves assault the berm, dramatically reducing its width, and in extreme cases virtually eliminating it (Figure 2-5). The sand that once composed the wide flat beach is then returned to the sea and deposited once again in offshore bars.

2-5. The berm is often obliterated during fall and winter.

The seaward edge of the berm is the dividing line between fore-shore and backshore, a boundary known as the *berm crest* marked by a distinct seaward slope towards the sea. Under the right conditions, waves may cut into the berm to such an extent that the crest takes the form of a distinctive vertical *scarp* a meter or more in height (Figure 2-6).

2-6. Carefully exposing the face of a scarp reveals its construction of successive layers of sand.

The Dunes

The most landward of the four coastal zones found in Florida is that of the dune. Dunes are defined as areas of permanent vegetation that are strongly influenced by the winds and spray of the ocean (Figure 2-7). These features are formed through the accumulation of wind-blown sand trapped by resident plants, resulting in the familiar border of rolling green hammocks flanking the shore. It is not uncommon for new plant growth at the dune margin to invade the backshore during the calm summer months. This apparent seaward progress of dune vegetation is generally reversed during winter

2-7. Dune plants are shaped by the sea wind.

storms, however, and only plants rooted beyond the destructive forces of this yearly onslaught persist from year to year.

COASTAL PROCESSES: SHAPING THE EDGE OF THE SEA

Beaches are transitional not only in the sense that they join land and sea; their very existence is transitional. They represent the ever-changing outcome of a set of complex and dynamic interactions; forces of wind and water eternally shaping and reshaping coastal environments. Beachcombing is enhanced then not only by knowing what sorts of things live on and near the seashore, but also by becoming aware of the ceaseless rhythms of the sea, for these create beaches, and are in the blood of every sea creature.

Waves

Wave-watching is possibly the most common beach pastime, a mesmerizing if not hypnotic experience (Figure 2-8). Beach-

2-8. Wave watching is a popular beach pastime.

combers are well aware of the fact that the sea never looks exactly the same on different days, no matter how many times we visit. A number of factors contribute to this result—lighting, water color, sky color, time of day, and season. But perhaps the most notable reason for the ever-changing face of the sea is the endless variety in the pattern of waves upon her surface. To begin to understand the countenance of the sea, it is useful to gain an appreciation for the complexity of the processes that lead to the birth, propagation, and death of ocean waves.

An ocean wave contains a great deal of energy—this much is obvious to anyone who has tried to stand in the path of a breaking wave. It is also quite apparent that larger waves contain more energy than smaller ones. The first question that we might ask of ocean waves then is *"Where does all that energy come from?"* The immediate source of ocean wave energy is the most obvious—the wind. The familiar waves we observe every time we visit the sea, no matter what the shape, or how large or small, are born when winds blow across the surface of the sea. The reason for this is conceptually simple enough—some of the energy contained in the moving air is transferred by friction to the water below. Describing exactly how this happens is not so simple, however, and oceanographers have written volumes on the subject, complete with mathematical formulas detailing the subtleties of the forces involved. Such elaboration is well beyond the intent and scope of this book, and readers with interests at such a level are referred to the many excellent works on basic oceanography available at most libraries.

An interesting aspect of wave generation not so obvious is the ultimate source of the energy contained in waves. The immediate source may be the wind, but what drives the wind? The answer is the heat contained in solar energy—particularly its uneven distribution upon the atmosphere of an inclined and rotating spherical planet. Thus, the power contained in ocean waves is generated in the sun—the same force that also often motivates our visits to the beach.

Let us consider now the basic anatomy of an ocean wave (Figure 2-9). Although oceanographers have a somewhat extensive set of terms available with which to describe the many aspects of waves, we will simply consider a few of the most fundamental properties. Wind-driven waves are not produced singly, but in sets

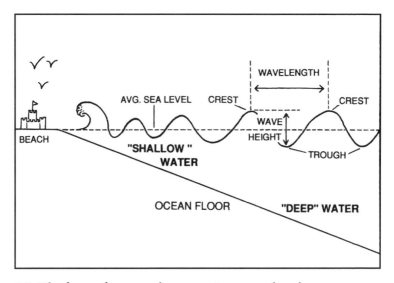

2-9. The form of a wave changes as it approaches shore.

with a good deal of variability in their individual properties. As they travel across the sea, trains of ocean waves are characterized by alternating high and low points, respectively, called *crests* and *troughs*. The distance between two successive crests (or troughs) is called the *wave length,* while the vertical distance from trough to adjacent crest is called the *wave height.* The *velocity,* or speed with which the wave moves is another key element in understanding waves, although somewhat more complex than wave length or height. This is because the speed at which the wave crest moves through the water, the speed at which energy advances in a wave train, and the speed at which individual particles of water move within a wave are all different yet related quantities.

As might be suspected, the size of waves generated by air moving across the water's surface is in part a function of the strength of the wind—the stronger the wind, the higher the waves. But two other factors also play a major role in determining the size and other properties of the waves ultimately generated. The first of these is the length of time the wind blows. Wave size will generally increase the longer the sea is exposed to a given wind, but only up to a point. Eventually, a maximal wave height is reached for any particular wind speed, and beyond that the waves will not increase regardless of the duration of the wind. Most (nearly 90%) of maximal wave development is reached in the first 24 hours of wind, with an additional 2–4 days required to complete the other 10%.

The third major factor affecting wave generation is called *fetch,* the unbroken distance of water over which the wind blows. Within limits, the greater the fetch, the larger the wave that may develop for a given wind speed and duration. To provide some common examples of these relationships, a fetch of about 200 miles is required to fully develop waves generated by a 20-knot wind. For a 40-knot wind, this distance increases to about 500 miles. Lakes, ponds, and other bodies of water with much more limited fetch never develop the really large waves seen in the open sea, regardless of the speed and duration of winds to which they may be exposed.

Most of the waves that we see breaking along the shore actually originate in strong storms far out to sea. Such storms produce a variety (spectrum) of waves, with differing speeds, sizes, shapes, and wave lengths. As these wave trains first move out from the center of the disturbance, they are generally in what is considered "deep" water—defined in this context as being greater than ½ the wave length. In a deep-water environment, the wave is allowed to roll on without interference from the seafloor. The average wave length of storm waves in the North Atlantic generally ranges from about 300–500 ft, so they remain in a deep-water environment until well up on the continental shelf.

A wave begins to interact with the seafloor below it as it first moves out of a "deep-water" environment and encounters depths less than ½ the wave length. This interactive process is formally called *shoaling,* but is also frequently and less academically referred to as "feeling the bottom." The possible interactions are numerous and complex, depending upon the angle at which the wave approaches the shore, the profile of the seafloor, the shape of the shore, and the characteristics of the wave. Because all these factors continually vary over time, even at the same beach, it is no wonder that the wave patterns we observe from the beach always look different.

Let us consider the transformation of a deep-water wave into a breaker as it approaches a beach (Figure 2-9). As the wave enters shallow water, it slows down, and the wave length and height decrease. But as the wave continues to near the shore, the height begins to increase and the wave eventually becomes higher than its original "deep water" ancestor. Depending upon the wave characteristics and the slope of the seafloor near the beach, waves generally take one of four basic forms to end their journeys. A *spilling breaker* breaks somewhat gradually as it proceeds shoreward, beginning at its top or crest (Figure 2-10a). This type of breaker forms when the bottom is rising quite gently at the point of the break. If the bottom slope is somewhat steeper, the wave may form a *plunging breaker* in which the face of the wave curls into a hollow tube enclosing a barrel of air (Figure 2-10b). The famous wave featured in the TV show *Hawaii Five-O* is a classic

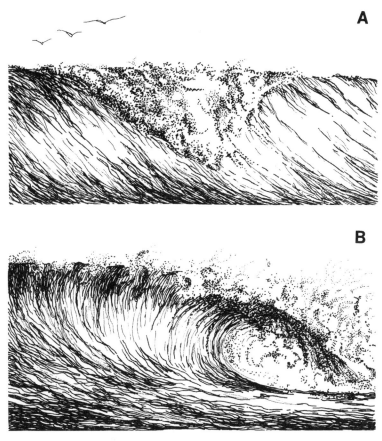

2-10. *Different seafloor characteristics result in different types of waves. Spilling (A) and plunging (B) breakers are the most common.*

example of the plunging breaker. If the slope of the seafloor is even steeper, the wave may collapse and only partially expend itself; the remainder of the energy contained in this *collapsing breaker* is reflected from the beach as a smaller wave moving in a seaward direction. Finally, on extremely steep beaches or sea walls, the *surging wave* may not break at all. Instead, it slides smoothly up the shore or wall for a short distance and then retreats, with much of its energy reflected back into the sea. When a wave in fact breaks, the energy it gathered so long ago and carried so far is quickly released and dissipated in the resulting turbulence. If it is reflected, it will move seaward until it meets the next shoreward-moving wave. This results in a phenomenon known as a *standing wave,* formed by the collision of incoming and outgoing wave trains. Standing waves sometimes result in spectacular "explosions" of crashing foam and are themselves often responsible for the creation of bars and longshore currents.

Tides

The movement of the tides has a major effect upon beaches and seashore life. The foreshore narrows and widens, ridges alternately become exposed and submerged, and longshore currents wax and wane as the line of breakers shifts shoreward then seaward. The tides also influence the movements, reproductive activities and other behaviors of many seashore creatures, as we shall see in later chapters.

Tides are the result of two opposing forces—gravity and centrifugal force—acting upon the earth's oceans. Let us consider a simplified model of the generation of tides to gain an understanding of some of the fundamental principles involved, keeping in mind that in reality the situation is much more complex. Imagine for a moment that the earth is a perfectly smooth solid sphere covered entirely by water, and that the seabed is frictionless with respect to the water above it. We will further simplify our model by assuming that the moon remains at a fixed distance above the earth, completing a perfectly circular orbit every 28 days. Finally, we will assume that the sun has no influence on our model. The question we will ask of our model universe is, *"Over time, how would the sea level change at different points on the earth's surface under such conditions?"*

First, we know that the moon would exert a gravitational pull upon the earth's watery shell, tending to pull the oceans towards the moon's center. The nature of gravity is such that the closer any point of the ocean was to the moon, the greater would be the pull of the moon's gravity at that point.

Counterbalancing this gravitational attraction would be a centrifugal force caused by the rotation of the moon and earth about one another. This force exerts its influence on the seas in a direction exactly opposite the gravitational pull of the moon. As a rough analogy, consider how centrifugal force causes the hair of two ice skaters to be continually thrown backward (outward) from their center of rotation as they spin about one another, held together by locked hands rather than gravity. However, unlike the gravitational pull of the moon, the strength of this centrifugal force remains the same at all points on the earth's surface.

If we assume that the two forces are of equal strength at all points on the earth's surface midway between that point on our planet closest to the moon and the point farthest away, we have all the information we need to answer the question posed. On a planetary scale, the net effect would be to deform the earth's watery envelope oceans from a spherical into somewhat of an ovoid shape (Figure 2-11). This would happen because except where the two

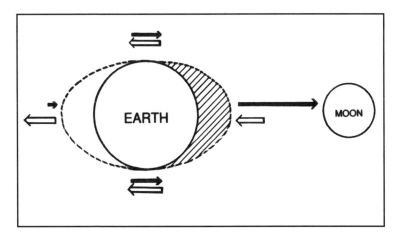

2-11. The gravitational attraction of the moon (dark arrows) and centrifugal force (light arrows) combine to create slight bulges in the ocean (greatly exaggerated here). These point directly toward, and away from, the moon.

forces were exactly balanced, one or the other would predominate, creating a slight bulge in the ocean in the direction of the prevalent force. On the side of the earth facing the moon, gravity would be stronger than centrifugal force, resulting on a bulge facing *directly towards* the moon. Simultaneously, at points on the opposite side of the earth, centrifugal force would overbalance gravity, creating a bulge in the sea facing *directly away from* the moon.

On a local scale, the combined effects of the two opposing forces—gravity and centrifugal force—upon sea level at any point on the earth's surface would depend upon their relative strengths at that particular point. Hence, the increased sea level represented by the tidal bulges would gradually lessen as one approached the line at which the two forces balanced, resulting in a belt of lower (relative to the bulges) sea level encircling the earth along that line.

If we now imagine the frictionless earth rotating smoothly beneath this watery shell every 24 hours, alternating areas of higher and lower water would pass over any given fixed point on the earth's surface, resulting in two equally high and low tides each day, with a separation of about 12 hours and 25 minutes between successive highs (or lows). This entire cycle of 24 hours and 50 minutes would correspond to the 24 hours it takes the earth to rotate once beneath the tidal bulges, plus 50 minutes to account for the movement of the moon in its orbit about the earth—a movement equaling 1/28th of the complete orbit.

Of course, reality is a good deal more complicated than our hypothetical model. The earth has an irregular surface, with land masses and ocean basins of differing shapes and dimensions. The moon has an elliptical rather than circular orbit around the earth, and the plane of that orbit does not remain fixed with respect to the Earth's equator. Many other complicating factors also exist, resulting in a much more complicated and geographically variable pattern of tides than our simple model would predict. Nonetheless, our model does a reasonably good job of accounting for one of the more common patterns of tides observed (semidiurnal). However, it does not explain the monthly tidal fluctuations—the so-called "spring" and "neap" tides. To do that, we must add in the effects of the sun.

The sun influences the tides in much the same manner as does the moon, although to a lesser extent. Because of its much greater distance from the earth, the gravitational pull of the sun upon the oceans is slightly less than half that of the much smaller moon. Because the sun, earth, and moon are constantly changing relative positions, the tidal effects of the sun are alternately adding to or subtracting from that of the moon. When the sun and moon are aligned with the earth, their gravitational effects are additive (Figure 2-12a), and tend to maximize the deformity in the shape of the oceans. This results in the most extreme high and low waters of the month, called the *spring tide*. The spring tide is highest (and lowest) when the sun and moon are on the same side of the earth, and is somewhat less when those two bodies are on opposite sides of the earth.

When the sun and moon form a right angle with respect to the earth (Figure 2-12b), their effects are antagonistic, resulting in the

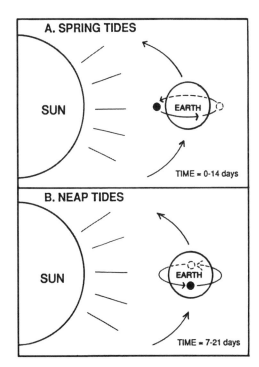

2-12. Spring tides (A) occur when the earth, sun, and moon are more or less aligned. Neap tides (B) occur when a line through the center of these bodies form a right angle.

monthly minimal or *neap tide*. Intermediate tides occur between spring and neap tides, when the sun and moon form angles between 0 and 90° with the earth. Successive spring and neap tides are separated by 7-day intervals, and there are two spring tides and two neap tides during each of the moon's 28-day orbits about the earth.

A number of differing patterns of tides characterize different coastlines of the world. In Florida alone, there are three broadly different tide regions. Tides along the eastern (Atlantic) coast of Florida in fact agree quite well with those that might be predicted by the model just discussed—there are two high and two low tides each day, of about equal amplitude. Successive high (or low) tides are separated by about 12 hours and 25 minutes, and the complete cycle of two highs and two lows is completed in 24 hours and 50 minutes. This basic tide pattern is called *semidiurnal*. In contrast, Florida's panhandle and some of the northwest coast of Florida (Gulf of Mexico) experience but a single high and low tide each day, separated by about 12 hours and 25 minutes, a *diurnal* tide pattern. The remainder of Florida—the west central to southwestern coasts and the southern coast (including the Florida Keys) are subjected to two unequal highs and lows each day, a pattern referred to as *mixed* tides.

Nearshore Currents

Currents in the coastal zone may result from both waves and tides, and along with these forces play a major role in structuring the edge of the sea. Waves breaking along a sandy coast may generate longshore currents that transport water and suspended sediments roughly parallel to the coast. Such currents are formed as incoming waves deposit masses of water into the troughs between offshore bars (Figure 2-13a). The water then flows along these troughs until it finds a way back out to sea. Longshore currents are also generated when waves strike the coast at an angle.

Sand bars of the surf zone are themselves transected at intervals by another set of channels, oriented more or less perpendicular to the shore. The longshore currents use these depressions, called *rip channels,* to return their waters seaward. The flow of water away from the shore in a rip channel is called a *rip current,* a phenomenon that is capable of placing unsuspecting swimmers at risk. Although it is a comparatively rare event to find oneself caught in a dangerously strong rip current, it is well to be aware of their existence and to know the proper steps to take if you or someone nearby does encounter this situation. In reality, it is not the rip current itself that kills most victims; rather it is the panic induced as the person begins to be swept seaward by a force too strong to directly resist. Fortunately, although they are often quite strong, rip channels and their rip currents are generally relatively narrow. Therefore, rather than fighting a losing battle to swim directly against the current, the proper techniques are to either swim parallel to the shore until you remove yourself from the current, or just try to relax and float with the current until it spans the bar and dissipates, which is usually quite quickly. At that point, it is relatively easy to swim along the beach until you are beyond the reach of the outflowing rip. Then, the incoming waves will assist your return to dry land. Although the experience of being caught in a rip current may be unnerving, be reassured that the vast majority of swimmers survive the experience. If you are ever caught, just try to remember that panic, not the current, is your worst enemy—a rip current will **not** continue to sweep you out to sea until you disappear kicking and screaming over the horizon!

Currents generated directly by tidal forces may also play a major role in shaping coastlines by distributing sediments. But tidally-generated currents only attain substantial speed and strength in places where the open sea communicates with other connected bodies of water—bays, rivers or lagoons—through restricted channels. The shoreward flow of water created by a rising tide near the mouths of inlets and estuaries is referred to as a *flood current,* while the corresponding seaward flow that accompanies a falling tide is termed an *ebb current* (Figure 2-13b). Flood and ebb currents attain maximal velocity midway between high and low water.

Compared with the role of wave-generated nearshore currents then, tidally-generated nearshore currents tend to play a comparatively restricted and localized role in the shaping of coastlines. Nevertheless, through their *interactions* with wave-generated currents, the tides also have a marked effect on the structuring of open coastlines. They do this by changing the sea level on a daily basis, thereby spreading the effects of wave-generated nearshore currents and other processes over a much wider area of the inshore zone than would otherwise be the case.

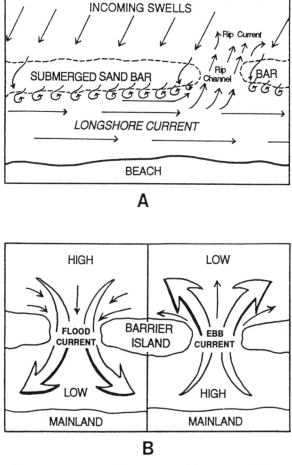

2-13. *Inshore currents may be generated by waves [A] or tides [B].*

Storms

Although severe storms are infrequent along Florida's shores, they have the capacity to rapidly alter sizeable stretches of coastline. Florida's beaches are subject to two major types of extreme weather systems with such effects. Tropical disturbances ranging from mild depressions to powerful hurricanes threaten the coast annually during summer and early fall, with peak activity from mid-August through mid-September. The strongest of these may pack sustained winds in excess of 200 miles per hour, and measure hundreds of miles across. The giant waves raised by such storms may in a few hours completely obliterate the berm built over an entire summer season and deposit its sands in offshore bars. The destructive "storm surge" at the eye of a hurricane may inundate primary and secondary dunes, cut new channels across barrier islands, and generally wreak havoc along miles of coastline. Fortunately, direct hits upon the Florida coast by storms of this magnitude are extremely rare.

The second type of seasonal weather system strongly affecting Florida's beaches is the so-called "norther" or "northeaster" of the fall and winter seasons. These storms are associated with the rapid passage of cold fronts sweeping southward out of the Canadian arctic, and bringing with them several days of strong northerly winds. The waves and longshore currents generated by northers have the capacity to cause massive beach erosion and sand transport.

Hurricanes and northers are parts of long-standing weather patterns that have built and altered Florida's coastline for many thousands of years. The changes wrought by these events may sometimes be catastrophic to the seashore life of limited areas. Nonetheless, these storms are also responsible for the creation of new beaches in other locations. Left to themselves, Florida's beaches and their living communities have a great deal of resiliency to such natural disturbances, and an ability to rapidly recover from their ravages. But poorly planned coastal construction has the capacity to magnify and alter the destructive effects of violent storms by disrupting the natural process that tend to maintain beaches in their natural state.

3

Florida's Beaches: Geology and Biogeography

GEOLOGY

F lorida is a true child of the sea. There is ample evidence of her oceanic origins locked within the marine sediments that form the bulk of the state's geological deposits. For most of the earth's long history, the area we now know as the Sunshine State might have been better called the "Submerged State," first emerging as dry land a mere 25 million years ago. The state's close affinities with the sea are maintained today to a greater degree than perhaps any of the other continental United States. Most of Florida's major population centers, as well as a large segment of its commerce, are intimately linked with the coast. The famous year-round mild climate is a direct result of the ameliorating influence of the surrounding ocean, perhaps the most obvious example of the many ways in which coastal and oceanic processes affect Florida's ecology. In fact, the entire state might be justifiably considered a rather large coastal zone, with a maximal elevation of less than 400 feet and no place more than about 60 miles from the ocean's shore.

The geological history of Florida has a great deal to do with the structure of her modern coastal zone habitats and their ecology.

Surprising though it may seem, current evidence suggests that the basement rocks below the Sunshine State were once part of a giant supercontinent called Gondwanaland that existed before the Atlantic Ocean formed. At that time, Florida was joined to the part of Gondwanaland that eventually became North Africa. When the supercontinent broke up some 200–300 million years ago, Florida was carried westward as the newly-born Atlantic widened. Eventually, Florida collided and merged with the young North American continent somewhere in the area of what is now southern Georgia.

During the 25 million years of existence above the waves, the dimensions of the Florida peninsula have changed dramatically along with the changes in sea level that accompanied the advance and retreat of the polar ice caps. Today, some 60% of the Florida coast—over 600 miles—is composed of sandy shorelines (Figure 3-1). The only major exceptions are the so-called "Big Bend" area of the Gulf Coast, and the extreme southern tip of the peninsula—

3-1. *The distribution of sandy beaches and seashore life (see page 65) in Florida.*

two regions that differ from the rest of the Florida coastline in that they are bordered by unusually broad and shallow marine plat-forms with extremely mild seaward slopes. Waves of appreciable size therefore "feel the bottom" and break far from shore, creating a quiet, low-energy lagoon-like inshore environment, even along these otherwise open unobstructed coasts. This in turn results in shores that are not only nearly wave-free, but also composed of fine sediments—mud and silt—rather than sand. The development of mangrove forests and marshes are favored under such condi-tions, and such habitats prevail in these two locations (Plate 1).

Rocky shores are not common in Florida, but still occur to a limited extent in several locations along the east coast, particular-ly in the central and northern portions. Just south of St. Augustine for example, one may visit shores lined by extensive outcroppings of Coquina rock—a geological product produced largely from the remains of thousands of the abundant tiny clams that abound in the foreshore and surf zones (Figure 3-2). Farther south, along shorelines of the east central coast, one may see shorelines that include rocky formations composed of sediments shaped into reef-like structures by the activities of millions of tiny tube-build-

3-2. Rocky shores along the northeast coast south of St. Augustine.

ing worms. In extreme south Florida, the remains of ancient coral reefs provide solid shorelines throughout much of the Keys. But most of Florida's shoreline is nearly pure sand. The composition and life of the sandy beaches that characterize the shorelines of the remainder of the state vary somewhat according to region. The size and nature of the particles that form a beach are among its most distinctive and meaningful characteristics, both from a structural as well as ecological standpoint. Florida's beaches are primarily composed of two types of sand particles; quartz and calcium carbonate. Quartz sands are the product of the weathering of continental rocks. Particles are carried to the coast by rivers, and over the years become dispersed, sometimes far from their original "seafall," by longshore currents. Most of Florida's east coast beaches are made up of quartz sands that have been transported southwards—sometimes over great distances—in just this way.

Calcium carbonate sands, in contrast, are derived from the activities of inshore tropical marine plants and animals (Plate 2). Many tropical organisms extract calcium carbonate from seawater and incorporate it into shells or skeletons. When they die, often by being eaten and crushed, fragmented remains accumulate in shallow waters where they continue to be broken down and abraded by wave action as they are gradually carried shoreward, eventually to form the distinctive beaches of tropical coasts.

Beaches of the northern Florida are mainly quartz along both the Gulf and east coasts. Farther south however, Florida's beaches gradually incorporate more and more calcium carbonate sands, and in the Miami area calcium carbonates makes up almost half the beach. The southernmost beaches of the peninsula and the Keys are made up primarily or exclusively of calcium carbonate sands and shell fragments derived from the tropical animals and plants that dominate the area.

BARRIER ISLANDS

The story of Florida's beaches is really the story of her barrier islands. Most of the state's sandy shores do not occur on the mainland proper, but rather along the seaward margin of a bordering network of long, narrow sand-covered islands. Barrier islands are often separated from the mainland by brackish lagoons and wetlands inundated by seawater mixed with freshwater runoff from mainland rivers and streams (Figure 3-3). These estuarine areas communicate with the sea through a series of inlets.

3-3. Barrier islands are separated from the mainland by estuaries and wetlands.

Florida's barrier islands have been connected to the mainland by an extensive system of bridges and causeways. These may create a bottleneck to traffic during times of heavy usage, and it is recommended that the savvy beachcomber plan accordingly and avoid the "rush hours." Nothing is more irritating on a perfect Florida day than to prime oneself for sun and surf, happily head off towards the sparkling sands, and wind up stuck in a long line of cars filled with other frustrated beachgoers all trying to reach the same place at the same time.

Although the origins of particular barrier islands are in many cases unclear, several alternate theories have been proposed to explain the most common processes generally involved in the formation of such islands. In some cases, they are believed to have originated from mainland coastal features—elevated dune ridges that later became isolated from the mainland by inundation of lower backlying areas as sea level rose. In other cases, they are believed to have originated as bars that gradually emerged from the sea and were subsequently cut into series of longshore islands by large storms or the slower development of transecting channels—our inlets of today. Alternately, some barrier islands formed from spits generated by longshore transport of sand. These spits then later became cutoff from the mainland by channels created in much the same manner as with islands formed from emergent bars. In some cases, a combination of two or all three of these processes may have contributed to barrier island formation.

The system of barrier islands that characterizes most of the Eastern Seaboard of the United States extends well to the north of Florida, all the way to New England. The network of lagoons and low-lying areas landward of these islands has been engineered to form the Intracoastal Waterway, a navigable protected passage that reaches all the way from Massachusetts well into the Florida Keys, and that is accessible to small and moderately sized vessels (Figure 3-4).

3-4. The intracoastal waterway provides calm passage along the entire length of Florida's east coast.

A notable characteristic of many of Florida's barrier islands is that they tend to have a rather gentle offshore slope, resulting in an extensive series of bars and ridges, and a correspondingly broad surf zone (Figure 3-5). This makes them ideal shores for beachcombers, because the shifting tides alternately cover and expose broad areas, daily creating ample new opportunities to find something as yet undiscovered by others. The same characteristics also make many of Florida's beaches ideal for wading and swimming, particularly during low tides when incoming waves break across the outermost bars, expending their energy at safe distances from the beach.

3-5. The gentle seaward slope of barrier islands creates broad flat beaches, good swimming, and great beachcombing.

As one proceeds inland on a barrier island, a variety of different types of habitats may be seen. The primary dune is usually the highest elevation of the island, its vegetation trapping and accumulating much of the beach sand carried inland by the onshore sea breeze. Behind the primary dune is generally found a depression or trough, often followed by a series of secondary dunes. The lines of dunes provide some protection from the wind-borne salt and sand, allowing less tolerant plants to thrive in their lee. Backdune areas may contain dense scrub vegetation or maritime forests. The world of the dunes is explored at length in Chapter 4. Finally, where the inland shore slopes gently into the quiet lagoon, marsh grasses and intertidal mudflats form wetlands. These are neither land nor water, but rather an intermediate mixture of both (Figure 3-6). The wetlands and lagoons that bridge

3-6. Wetlands are productive habitats that connect land and sea.

Florida's barrier islands and mainland are incredibly rich and productive ecosystems in their own right. They serve not only as the primary habitat for estuarine life, but additionally represent a vital temporary habitat for many marine animals, including a number of commercially valuable species. These sea creatures depend upon wetlands and lagoons for such necessities as seasonal refuge, rich feeding grounds, and a safe place to reproduce and develop during their early and most vulnerable life stages. The continued health of Florida's coastal marine communities is closely and inexorably linked to the health of adjacent wetland and lagoon environments.

BIOGEOGRAPHY

Florida's Marine Invertebrates

Before describing the regional distribution patterns of Florida's marine fauna, it would be well to include a brief overview of the diverse groups of animals that we are discussing. Marine animals include both those with backbones—the *vertebrates*—and those without—the *invertebrates*. This section focuses on the latter, because most readers are much more familiar with the general classification, structure, and function of animals with backbones—mammals, birds, fishes, reptiles, and amphibians—than they are with the generally smaller, less conspicuous invertebrate animals.

Despite the admittedly biased view we humans hold regarding the supremacy of vertebrate animals, it is a somewhat humbling fact that of the 26 or so major divisions (phyla) recognized within the animal kingdom, only **one** contains animals with backbones—the other 25 are exclusively composed of invertebrates. The amazing variety of form, life histories, behavior, and ecology of marine invertebrates would fill many volumes. This section presents the general characteristics of the most commonly observed groups of invertebrate animals that make up Florida's inshore marine fauna, animals that we will be referring to again and again in subsequent chapters.

Sponges. The sponges (Plate 3) are a colorful group. They may be quite large, and some species are able to easily contain a scuba diver with full gear in the central cavity. Sponges are considered to be the most primitive of all animal groups, and some biologists have argued that they ought not to be considered animals at all, but instead placed in their own separate group. They are extremely simple in structure and function, and possesses no nervous system, digestive system, nor most of the other types of specialized tissues and organs that characterize more advanced animals. The sponge body is little more than a perforated sack through which seawater is filtered and tiny organisms and bits of organic matter are removed and digested. Special cells create tiny currents that funnel the surrounding seawater through the body wall and into a central cavity. In the process food is removed, and the water is then expelled through a large opening. The spacious body cavities of sponges form a habitat in themselves for a variety of fishes as well as other invertebrates.

Cnidarians. This group, previously known as the *coelenterates,* contains a number of familiar animals such as jellyfish (Plate 4), corals (Plate 5) and sea anemones (Plate 6), as well as some less familiar forms. Cnidarians have a body plan based upon radial symmetry, a primary characteristic of which is that the animals are approximately circular in outline. The group is also distinguished by the presence of specialized stinging cells called nematocysts that are used in defense as well as to capture prey.

Cnidarians occur in two basic body forms, respectively called the *polyp* and *medusa*. Despite outward appearances, the two forms are actually quite similar in structure and function. The medusa (Figure 3-7) is a free-swimming form in which the mouth

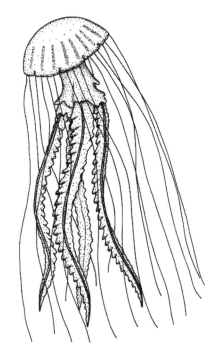

3-7. The medusa body form is typified by the jellyfish.

and tentacles face downward and are capped by a large muscular bell. Polyps in contrast are attached to solid substrates, and take the form of an inverted medusa with the mouth and tentacles facing upward from a muscular stalk or base (Figure 3-8). For most beachcombers, the most readily observed polyp form is the familiar sea anemone (Figure 3-9; Plate 6) often seen at low tide on solid substrates.

3-8. The polyp body form is adopted by corals and sea anemones.

3-9. Warty sea anemone (Bundodosoma cavernita).

Medusae exist as solitary individuals, whereas polyps often form large colonies composed of many thousands of genetically identical individuals, as most spectacularly illustrated by the reef-building corals that thrive in the warm clear waters of the Caribbean and the Florida Keys (Plate 5). Some kinds of cnidarians exhibit a life history pattern that includes both medusa and polyp forms, while others are exclusively composed of individuals of just one of these two possible body plans.

Polychaetes. Polychaetes are among the most advanced and active of worms. Their name literally means "many-bristled" and denotes one of the group's most characteristic traits. Some live out their adult lives as sedentary forms that produce protective tubes or burrows attached to the substrate, and from which filtering structures protrude to sweep passing plankton from the sea (Figure 3-10; Plate 7). But many of the mobile free-living polychaetes are

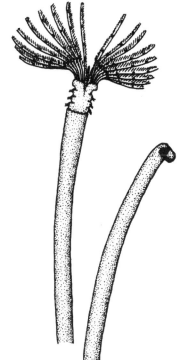

3-10. A sedentary polychaete shown with feeding tentacles extended (left) and withdrawn (right).

scavengers or voracious predators (Figure 3-11; Plate 8). Polychaetes are a major link in many marine food webs, and frequently represent a major food source for many kinds of fishes.

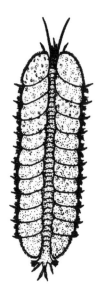

*3-11. Scaleworm (*Lepidonotus sublevis*)–a free-living polychaete.*

Bryozoans. These tiny animals are often mistaken for plants, as they grow in colonies that sometimes form a colorful film-like coating upon the exposed surfaces of rocks or larger animals and plants (Plate 9), or alternately take on the bush-like form of a small clump of seaweed (Figure 3-12). Individuals are encased in tiny protective shells that grow together in intricate and delicate patterns often too small to be really appreciated by the unaided eye. Bryozoans obtain food by straining tiny drifting organisms from the sea.

Echinoderms. The spiny-skinned animals, or echinoderms, include a variety of sea creatures that one would not necessarily recognize as close relatives. Sea stars (commonly called starfish) occupy a variety of seafloor habitats where they prey heavily upon shelled mollusks (Figure 3-13). Brittlestars are smaller and

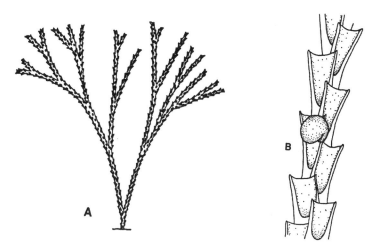

3-12. A bryozoan colony (A) with a close-up of individual members (B).

3-13. Sea star (Astropectin duplicatus x 0.3).

a good deal more lightly built than sea stars (Figure 3-14; Plate 10). They occur just about everywhere on and in the seafloor. Brittlestars are mainly nocturnal predators that remain sheltered by day, often buried in bottom sediments or within the cavities of sponges. They feed in a variety of ways, as scavengers, predators, and detritus feeders. Brittlestars are themselves frequently con-

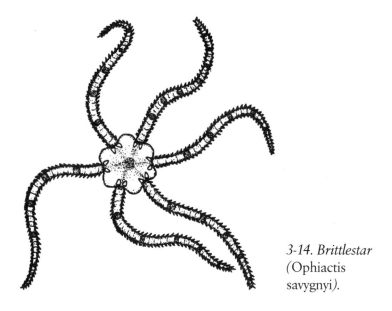

*3-14. Brittlestar
(*Ophiactis
savygnyi*).*

sumed in large quantities by fishes. The familiar sea urchins and their relatives (Plate 11) are active herbivores.

Mollusks. The mollusks are even more diverse than the echinoderms, with many different species, body plans, and ways of life. The first of the groups of particular interest here are the snails and their relatives. Some are slow-moving grazers on marine plants, while others are active predators on other animals.

Bivalve mollusks include the familiar clams, mussels, oysters, and scallops, and are characterized by a body encased in two plate-like shells, hinged along one side (Figure 3-15). Bivalves are active filter feeders, pumping water through specialized gills to remove food. Many are heavily hunted by fishes and other predators.

The active swimmers of the group—the squids and octopi—have well-developed nervous systems, complete with relatively large brains and eyes very much like our own. They are masters of color change, often so well camouflaged as to be virtually invisible even at close range. Octopi are benthic creatures (Figure 3-16) that may be occasionally located by beachcombers in rocky

(continued on page 63)

Plate 1. Mangroves line protected shores.

Plate 2. Parrotfish may produce a ton of sand each year by eating and crushing coral.

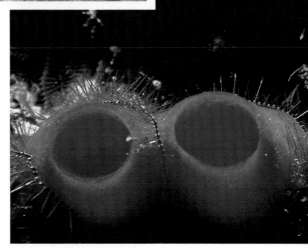

Plate 3. Sponges add color and structure to solid substrates.

Plate 4. Jellyfish are among the most primitive of animals.

Plate 5. Two types of coral: a boulder-shaped reef-building colony (background) and a delicate branched octocoral (foreground).

Plate 6. Pink-tipped sea anemone.

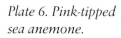

Plate 8. A predatory free-living polychaete feeds on a sea fan.

Plate 7. The featherduster is a sedentary polychaete.

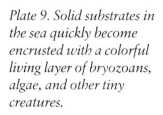

Plate 9. Solid substrates in the sea quickly become encrusted with a colorful living layer of bryozoans, algae, and other tiny creatures.

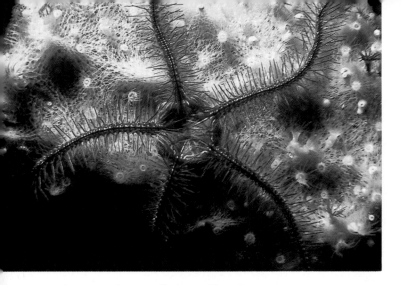

Plate 10. The spindly-legged brittlestar.

Plate 11. A sea urchin perches atop a coral colony.

Plate 12. Sea rocket (Cakile edentula).

Plate 13. Orach
(Atriplex arenaria).

Plate 14. Sea oats (Uniola peniculata) *rise along the crest of the primary dune.*

Plate 15. Sea grape
(Coccoloba uvifera).

Plate 16. Spiked seeds of
the sandbur (Cenchrus
tribuloides).

Plate 18. The beach morning
glory (Ipomea pes-caprae).

Plate 19. Seaside bean
(Canavalia maritima).

60

Plate 20. Saw palmetto
(Serenoa repens).

Plate 21. Cabbage palm
(Sabal palmetto).

Plate 22. Spanish bayonet
(Yucca alifolia).

61

Plate 23. Prickly pear cactus (Opuntia stricta).

Plate 24. Grasslands dominate the coastal vegetation along much of Florida's Gulf Coast.

Plate 25. A subtidal hermit crab.

(continued from page 54)

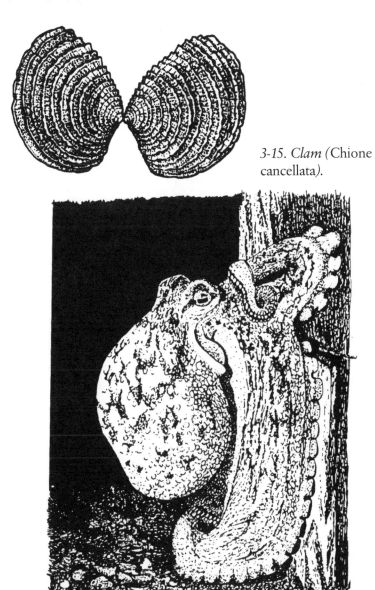

3-15. *Clam (*Chione cancellata*)*.

3-16. *The shy octopus inhabits shallow rocky shorelines.*

subtidal areas, while the squid (Figure 3-17) is an open water hunter that feeds on a variety of large prey, including fishes.

*3-17. Squid (*Loligo peali, *x 0.5).*

Crustaceans. The crustaceans include shrimps, crabs, lobsters, and a host of much smaller and less familiar relatives such as amphipods (Figure 3-18) and isopods (Figure 3-19). Surprisingly, the barnacles are also members of this large and diverse group.

Crustaceans play a host of different roles in the life of the sea. Some are scavengers, cleansing the seafloor of decaying animal remains. Others are active predators on larger animals or on tiny plankton. Some shrimp are "cleaners," gaining food by removing parasites from fish or invertebrates. All have a protective outer shell or skeleton, which, through a process called *molting,* must be periodically shed and replaced as the animal grows.

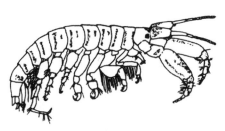

*3-18. Tube-building amphipod (*Corophium louisianum, *x 10).*

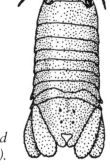

*3-19. Squat isopod (*Dynamene perforata, *x 10).*

Distribution of Marine Animals

Florida possesses three different kinds of regions when it comes to animal life of the seashore (Figure 3-1, see page 41). The northeast coast, extending from the Georgia border to around Cape Canaveral, represents the southernmost part of what biogeographers call the *Carolinian Faunal Province,* a subtropical area extending northward to Cape Hatteras. Because the animal assemblages of the northern Gulf of Mexico are highly similar to those of the northeast coast of Florida, the seashores of Florida's panhandle are generally considered part of a disjunct extension of the Carolinian Province.

The animal assemblages that characterize the beaches of south Florida and the Keys, as well as the islands and coastlines of the Caribbean Sea, belong to a truly tropical *West Indian* fauna that differs in distinct ways from the more northerly Carolinian fauna. There is not a sharp dividing line between the Carolinian and West Indian faunal provinces; rather, they gradually grade into one another. The beaches located between the two regions are therefore inhabited by a mixture of animals from both regions. Such an intermediate or mixed fauna occurs along Florida's east coast from about Cape Canaveral southward to Miami, and on the west coast from the Tampa Bay area southward to around Cape Romano. Superimposed upon these regional patterns is a "cosmopolitan" component, consisting of wide-ranging animals that have been transported to Florida by winds, currents, or as "hitchhikers" on boats or floating debris.

The reader is advised that these rather broad characterizations of the regions described should not be taken too literally. Depending upon currents, water temperatures, and season, it is not uncommon for what are generally considered strictly tropical (West Indian) animals to appear in the Carolinian Province, or even farther north, nurtured for a time by temporarily benign conditions. But such faunistic intrusions are not permanent. Inevitably, conditions once again become too harsh for truly tropical species, and they perish far from their rightful homes.

Seaside Plants

Not surprisingly, there are some differences between the way animal biologists (zoologists) and plant biologists (botanists) view the coastal biogeography of Florida. This should not be particularly worrisome to the reader. The distribution and abundance of land plants along the coast are controlled by different processes (i.e., soils, rainfall, etc.) from those responsible for the distribution and abundance of seashore animals. The latter are primarily the result of oceanic processes (i.e., waves, tides, currents, water temperature, seafloor characteristics, etc.).

Thus, based upon similarities and differences in coastal vegetation, plant specialists tend to recognize five distinct regions among the state's beaches, rather than the three faunal provincial characterizations of the zoologists (Figure 3-1). Beginning along the northwest coast, the first region extends from the Alabama border to the beginning of the "Big Bend." The second area begins just to the north of the Tampa Bay region and extends southward to around Cape Romano. Moving to the extreme southern portion of the state, our third region includes the beaches of Cape Sable and the Florida Keys. The fourth region extends just south of Miami northward to around Cape Canaveral, and the fifth and final region from Cape Canaveral to the Georgia border. Specific differences in the coastal plant communities inhabiting each of these regions will be discussed in the next chapter.

Despite certain differences, there is a clear relationship between the floristic regions described here and the faunistic regional characterization presented earlier. The panhandle and northeast coast floristic regions together comprise what we earlier referred to as two disjunct segments of a single Carolinian Faunal Province. The southernmost floristic area—Cape Sable through the Keys—corresponds nicely to the part of the state characterized by a true West Indian marine fauna. And finally, the southeast and southwest coastal floristic regions correspond to areas where the animal life is considered an intermediate mixture of Carolinian and West Indian components.

4

Dunes: Last Stop for Land Plants

The landward margin of most of Florida's beaches is rather sharply defined by a band of deep green foliage cloaking the adjoining dunes (Figure 4-1). Dunes interact with the beach prop-

4-1. The sharp dividing line between dune and beach is highlighted by the contrasting colors of open sand and dense vegetation.

er in many ways in terms of both physical and biological processes. Anyone who visits Florida's beaches on a regular basis soon becomes familiar with common dune plants, many of which are seen along the entire 1,000+-mile coastline. In this chapter, we explore the physical and biological processes involved in dune formation and evolution, as well as common plant and animal life of this most landward of the coastal zones.

THE EVOLUTION OF DUNES

The dunes that border most of Florida's beaches are really rather simple in structure, being little more than small sandy hills held together by the roots of covering vegetation. Nonetheless, the origins and subsequent evolution of coastal dunes are more complex than one might guess from their rather ordinary appearance, and understanding these processes and their implications is a vital aspect of coastal zone management and conservation. A consideration of how dunes are born and how they further develop is also central to understanding the common patterns of habitat and plant zonation that characterize Florida's many barrier islands.

To illustrate the essential elements of the processes involved, let us imagine a newly born barrier island, recently emerged from the sea. Our island is at first nothing more than a flat platform of barren sand lying just a few inches above the surrounding water, with the seaward shore exposed to incoming waves. Most of the energy contained in these waves is released as they break on the offshore bars. But as the wave remnants continue landward and finally strike the outer foreshore, there is enough remaining force to push a thin veil of water, called *swash,* up the beach (Figure 4-2). The swash carries with it tiny particles of suspended organic and inorganic debris of all sorts, which tends to be deposited as lines of *wrack* at the furthest reach of the swash.

Several lines of wrack roughly paralleling the shore are often evident. These result from wave trains of varying sizes arriving at different times and on different tides. Over time, the sea will reclaim much of the debris previously carried up onto the beach

4-2. Swash—the foamy result of the last remaining energy of ocean waves.

by waves, as successive high tides and storms reach farther and farther up the shore and sweep it clean. But what about the debris carried farthest up the beach by the largest storms and highest tides? This highest drift line is beyond the reach of lesser encroachments by the sea, and therefore remains undisturbed for long periods (Figure 4-3). It is this very material that often accounts for the genesis of dunes. Here, along with the inconsequential flotsam and jetsam, rich organic material is decaying and releasing nutrients that will soon nurture the germinating seeds of pioneering land plants. Some of the seeds may have washed ashore with the debris itself, while some are haphazardly deposited here by the wind.

As the first shoots emerge and roots develop, sand blowing ever shoreward by the sea breeze or occasional offshore storm becomes caught and trapped by the young plants. Tiny localized hillocks, the embryonic beginnings of a dune, soon form around the bases of the first clumps of vegetation (Figure 4-4). As sand continues to accumulate, the plants grow to keep pace. Nearby

4-3. Wrack is often the birthplace of the primary dune.

4-4. Emergent shoots begin to trap wind-blown sand.

hillocks now begin to overlap and lose their identities as they merge into an ever-larger structure forming just landward of the original drift line. And so in time, an elevated vegetation-clad ridge paralleling the beach begins to rise to face the sea. A *primary dune,* or *foredune,* has been established.

Occasionally, sand builds up faster than the plant grows, and the shoots are entirely buried. The hardy pioneers that first establish the primary dune are not deterred by such insult however; in fact, they are particularly adapted to not only survive but to actually grow their way right back into the sunlight again. As the dune continues to rise, its sand is gradually stabilized and enriched by the extensive network of interconnected roots of the overlying carpet of vegetation. Decaying plant matter, and the associated microorganisms that digest and process such material, release vital nutrients into their immediate surroundings. These then percolate and disperse through the loose sand, leading to the formation of a moist organic layer just below the surface of the growing dune. A ready supply of critical nutrients along with a means of protection from the full ravages of the sea wind and spray now allows less hardy plants to successfully colonize the crest and landward side of the dune, and a more complex plant community is thereby established.

The wind continues to move sand landward from both the berm and the dune face. Over time, the net result of this slow but inevitable process is to cause the dune to continue to grow both in height and width, and to eventually migrate inland. As the original dune retreats from the sea to become a *secondary* or *backdune,* primary dune formation continues at the place vacated—the highest drift line. Given enough time, the overall process results in the formation of a dune field—a series of dunes at different stages of evolution, with the youngest (foredunes) at the sea's margin and the oldest (backdunes) farthest inland (Figure 4-5).

Different environmental conditions exist in various parts of the dune field, allowing a diversity of plant species to become established. Dune vegetation is distributed in generally predictable patterns within the spatial context of the dune field. We will now

4-5. The deep furrow (left center) cleanly divides the primary dune (left) from the secondary dunes (right). Note the difference in the plant assemblages of each.

look at the most common dune plants and their distribution patterns along Florida's coastline.

PLANTS OF THE DUNE

Dune vegetation typically displays a zonation pattern consisting of parallel bands following the shoreline. Three broad areas are typically distinguished, based on the characteristic plant associations found in each. These are the foredune, the most exposed region; a transitional or intermediate zone consisting of a varied mixture of less tolerant plants; and farthest inland a relatively stable backdune inhabited by a more permanent, long-lived assemblage dominated by larger trees and shrubs.

Despite the rigors of beach life, a surprising number of plant species may be found along Florida's beaches. Just over twenty kinds of plants are *endemic* to Florida's beaches, occurring naturally nowhere else. If we were to compare detailed censuses of plant species found along transects leading from the beach through the backdune in the five biogeographic regions described in Chapter 3, we would find some notable differences among regions. We might also find substantial differences among different sites *within* the same region. Such variability is to be expected in most biotic communities, because the abundance and distribution patterns of the many species comprising a community are the products of the unique genetic makeup of each species interacting with a wide array of environmental variables and chance factors. This leads to a high degree of variability in the precise structure of complex communities, with predictable patterns only discernable on a relatively large spatial scale and in general terms.

For these reasons, we will not attempt to list or describe each of the hundreds of different kinds of plants one *might* see bordering Florida's beaches—there are other books that fill that need. Instead, we will focus here on plants that the Florida beachgoer will *probably* see again and again in most locations—the common dominant elements of each of the three zones comprising Florida's coastal plant communities.

Regional Variation

Florida's coastal plants exist in a transitional zone between the warm temperate region just to the north and the tropics to the south. As with many of the marine animals, this major climatological shift is reflected in the distribution of the state's beach flora. Tropical animals and plants are simply not adapted to withstand severe cold or hard freezes, even for relatively brief periods. In practical terms, this requirement limits truly tropical species to areas southward of places in Florida where such rare winter events are likely to occur. This invisible and diffuse boundary lies in the vicinity of Cape Canaveral on the east coast and Tampa Bay on the west coast.

A temperature-mediated distributional effect is less pronounced among plants of the foredune than those of the two more landward vegetation zones. The reason is simply that plants that are closest and directly exposed to the sea are somewhat more insulated from sudden atmospheric temperature changes than those not so positioned. When cold air descends quickly upon the Sunshine State from the frozen arctic wastes, air in close contact with the sea surface quickly picks up the water's heat. This creates, close to the sea, a layer of air of more moderate temperature than that of the overlying air mass. Evidence of the degree of protection gained by proximity to the sea may be seen by the fact that over one-third of all beach/foredune plants are able to range throughout the state, whereas only around 15% of plants found in the intermediate and backdune zones are capable of this feat.

Plants of the Foredune

The dune that immediately faces the beach is the harshest of environments for land plants, particularly on the seaward face. Plants here must deal with the stress of being alternately bathed in ocean spray and rinsed with fresh rain water. They live in an area often exposed to sustained high winds that are capable of quickly burying the entire plant beneath a blanket of sand. Visible evidence of the constant effects of the sea wind upon the face of the dune is frequently present in the form of intricate rippled sand patterns etched there (Figure 4-6). Wind also acts in combination with intense sunlight to draw vital moisture from the plant body. Occasionally, high tides and storms produce waves that actually rip into the dune face, tearing away sections of sand and with it the vulnerable vegetation.

Survival on the exposed foredune therefore requires some special adaptations not necessary in more benign environs. A high tolerance for salt and an ability to conserve water are two of the most immediate requirements. Water is conserved by plants of the primary dune through many of the same mechanisms used by desert plants. A greatly reduced surface area available for evaporation is one such widespread adaptation. Because many broad flat leaves present a large surface area, plants adapted to unusually dry conditions tend to greatly reduce both the number of leaves

4-6. Wind-carved ripples in the face of the primary dune.

and their surface area. This translates into thick compact leaves in which water may be more readily retained in the face of heat and drying winds (Figure 4-7). Cacti (Figure 4-8) are the most extreme example of this strategy, doing away with conventional leaves completely and relying instead on an expanded and specialized stem to absorb the sunlight needed for photosynthesis.

Another common adaptation seen in plants of the foredune for minimizing water loss is to cover vulnerable surfaces with an unusually thick and water-resistant waxy outer covering called a *cuticle*, giving a brilliant glossy lustre to the leaves of some dune plants (Figure 4-9). Seeds must also be unusually water resistant, and many are capable of surviving an extended trip in the sea to reach shore and germinate at a far distant beach.

4-7. Short, thick leaves are an effective adaptation for conserving water.

4-8. The prickly pear cactus is a common plant of American deserts and Florida dunes.

4-9. Many beach colonizers have shiny leaves, denoting the presence of a thick protective coating that serves to minimize moisture loss.

The same adaptations that reduce water loss—reduced surface area and thick insulating cuticles—also serve to minimize salt absorption from the environment in these plants. Nonetheless, constant exposure to sea spray leaves its mark, and plants of the foredune have necessarily developed an unusual tolerance for high concentrations of salt in their cells and tissues.

Another adaptive ploy of beachfront plants is to adopt the low-profile growth form of vines and creepers, presenting less resistance and exposure to the assaulting wind and the sea spray it carries (Figure 4-10). Whatever the growth form, plants of the foredune must be able to anchor themselves firmly to the sand below early in their lives, or surely be swept away. This goal is most frequently accomplished by a dense network of interwoven roots that grow and spread quickly during early developmental stages.

Many of the most common plants inhabiting Florida's primary dunes are considered true pioneers, often successfully colonizing parts of the beach itself. One such common inhabitant of the foredune face is the sea rocket, *Cakile spp.*, a dark green herb with

4-10. Low creeping vines offer minimal resistance to the sea wind.

small purplish flowers bearing four petals (Plate 12). The leaves of the sea rocket are thick and narrow, with irregular scalloped or toothed edges. Flowers are generally visible only during fall.

Two other herbs accompany the sea rocket as the earliest of dune builders. These are the orach (*Atriplex pentandra.*), a grayish plant with long arrow-shaped leaves (Plate 13), and the beach croton (*Croton punctatus*), a pale green plant with thick oval leaves of fine fuzzy texture. All three of these pioneering herbs range throughout Florida and well to the north, with the beach croton common through North Carolina and the other two reaching all the way to New England.

Joining the herbaceous early colonizers are several grasses. The most ubiquitous and well-known of Florida's foredune grasses are the lovely sea oats, *Uniola paniculata* (Figure 4-11; Plate 14). The sight of this tall tan grass swaying gracefully in the sea breeze has become more or less symbolic of Florida's beaches. Sea oats were for many years heavily collected for use in dried flower arrangements. Today, in recognition of their vulnerability and their pivotal

4-11. Waving sea oats are a familiar sight along the Florida coast.

role in dune formation and stabilization, sea oats are protected by law and should be seen but not otherwise disturbed.

Sea oats are the dominant foredune stabilizers along the state's northern and east central beaches, but farther south the distinctive sea grape (*Coccoloba uvifera,* Plate 15), becomes increasingly prominent in this role. The sea grape is also protected by law, and must not be collected or damaged.

The sandbur (*Cenchrus tribuloides,* Plate 16) is also a pioneering grass that is frequently found growing amid sea oats on the foredune, although it also is common on the protected backside of the foredune and on secondary dunes. This grass is green during early summer, but gradually lightens to a tan color as the summer progresses. It may grow to well over a foot, and bears multi-spiked burs that are a formidable deterrent to unprotected visitors.

Picking the burs of this plant out of one's socks is a regular part of visiting beaches from Florida to New York. The sandbur also accounts for much of the frequent hopping and one-legged stances characteristic of those unwise enough to attempt to venture into Florida's dunes barefoot. Another common dune grass is *Panacium amarum*, commonly called beachgrass.

As the primary dune forms and becomes stabilized, other colonists eagerly take up their places in the new habitat created by the first pioneers. One such early arrival is the seaside purslane (*Sesuvium portulacastrum*, Plate 17), a tropical herb that ranges from Florida southward through the islands of the Caribbean and along the Atlantic coast of Central America. This plant grows as a low dense mat composed of greenish or reddish stems and narrow bright green leaves.

Several vines are also common and widespread where dune and beach meet. The beach morning glory (*Ipomoea pes-caprae*, Plate 18) grows close to the sand, extending runners across the lower dune face and onto the beach for as far as the sea will permit. Single plants may extend 20 feet or more across. Similar in general form and appearance to the beach morning glory, but usually somewhat smaller, is the seaside bean (*Canavalia maritima*, Plate 19). Both vines have shiny bright green leaves and purple flowers that contrast vividly with the whiteness of the surrounding sands, and both flower throughout the year.

Plants of the Transitional Zone

The secondary dunes represent a somewhat more protected and enriched environment for plant recruitment and growth than the less hospitable foredune. The zone begins just behind the protecting barrier of the primary dune, and extends to the area in which a relatively stable coastal plant community, often including taller trees, is established (Figure 4-12). With a measure of protection from wind and sea, and access to soils already stabilized and enriched by the rugged pioneers that originally built the dune, other plants are now able to not only gain a foothold, but even outcompete the pioneering species so beautifully adapted for harsher

4-12. Low trees and shrubs of the intermediate or secondary dune give way to a maritime forest, (St. John's County, northeastern Florida).

conditions. Nonetheless, this zone clearly still experiences a sufficient level of stress from the effects of the sea wind and shifting sands so as to preclude colonization by the more stable but less tolerant plant assemblages common to more inland areas.

The transitional zone also contains the greatest number of coastal plants, including the majority of endemic species. In part, this high species diversity may be attributed to the fact that the physical structure and habitat characteristics of this zone are the most variable of the three. There are also more distinct systematic regional differences in the plant assemblages found in this zone than those of the foredune.

The transitional zone in most areas tends to be dominated by low growing woody shrubs intermixed with dwarfed growth forms of trees more common to inland plant communities. The vegetation of this zone is also referred to in some literature as coastal *strand,* a term that is sometimes used to include the plants of the foredune as well.

Along the northeast coast (Jacksonville to Cape Canaveral), the seaward segment of the transitional zone is often largely composed of the saw palmetto (*Serenoa repens,* Figure 4-13, Plate 20), found in dense stands just behind the primary dune. This plant extends a cluster of formidable spike-like leaves three feet or more in length. Farther inland, saw palmetto is gradually replaced by a plant assemblage referred to as *coastal scrub.* This includes a variety of small shrubs and commonly a dwarfed growth form of the live oak (*Quercus virginiana*), along with similarly modified forms of several other trees common to more inland coastal forests. The reduced size and unusual shapes of these shrubs and trees is a direct response to constant exposure to the sea wind. Common members of this assemblage include the cabbage palm (*Sabal palmetto,* Figure 4-14, Plate 21), red bay (*Persia borbonia*), wax myrtle (*Myrica cerifera*), and yaupon (*Ilex vomitoria*).

Transitional zones of the southeast coast (Cape Canaveral to Miami) are somewhat different. Cape Canaveral itself is some-

4-13. Saw palmetto sometimes occurs in dense stands in the seaward portion of the secondary dune.

*4-14. The cabbage palm (*Sabal palmetto*) is found throughout Florida in both coastal and inland habitats.*

what unusual in that its low broad expanse behind the foredune is covered mainly by grasses (*Muhlenbergia capillaris* and *Andropogon* spp.). South of the Cape, woody species once again dominate the transitional zone. However, now tropical shrubs begin to mix with the more northerly warm temperate species, and saw palmetto and dwarfed live oak are increasingly replaced by the sea grape.

An inhospitable (for beachcombers) band of well-defended plants frequently separates the stands of sea grape from the vegetation of the foredune along the southeast coast. This so-called "prickly zone" contains most notably Spanish bayonet (*Yucca alifolia*, Plate 22) and prickly pear cactus (*Opuntia stricta*, Plate 23).

The southwest region (Cape Sable to Tampa Bay) has a very different transitional zone than that of the east coast, probably due in a large part to the absence of a prevailing sea breeze during much of the year. Here, the sea grape is joined by a variety of other tropical shrubs such as the nickerbean (*Caesalpinia bonduc*) just behind the foredune. Proceeding inland, the shrubs tend to rather quickly give way to grasslands, rather than the dwarfed trees and shrubs that characterize the same zone along the east coast of the state (Plate 24).

The transitional zone of the coast along Florida's panhandle is, like the southwest coast, also largely dominated by grasses rather than trees and shrubs. These are endemic to the northern Gulf Coast region however, and differ from those species most common along the southwest coast. The predominance of grasses and absence of maritime forests along the Gulf coast may seem somewhat surprising, considering the strong general climatological similarities between the panhandle and the northeast coast of the state. However, the Gulf coast lacks the strong prevailing onshore breezes that characterize Florida's east coast, a difference that has far reaching effects on the structure and vegetation of the dunes.

Plants of the Stable Backdune

Environmental conditions in this most inland of the three coastal vegetation zones are such that stable mature communities are able to persist. In many cases, these same plants are found far inland, but those near the sea are clearly still subject to somewhat harsher environmental conditions. Along the east coast, particularly in the northern portion of the state, the plant communities found here take the form of maritime forests, with a canopy formed by the taller trees and an understory constructed of smaller, more shade-tolerant trees and shrubs (Figure 4-15). In other cases, the backdune community is composed primarily of low scrub vegetation or grasslands, or a mixture of the two. Well-developed maritime forests are uncommon along the coast of the panhandle. They are much more prevalent along the east coast of Florida, while grassy savannahs and scrub are characteristic of the west and northwest coasts.

4-15. *The mature maritime forest.*

The trees and shrubs found in the stable backdune include many of the same species also found, albeit in modified form, in the transitional zone. Although they assume growth forms more typical of the species, plants of the backdune are not completely immune to the sea wind and other harsh aspects of life at the edge of the continent. The tops of trees are frequently flattened, as though sheared off by the prevailing breeze and its saline cargo. Many trees and shrubs assume an asymmetrical form in response to the stress of the wind on their exposed quarters.

Backdune areas of the northeast region are largely vegetated by a community type known as live oak forest, dominated by the namesake of the community *Quercus virginiana,* but a number of other species also are characteristic of the region and join with live oak to form the forest canopy. These include laurel oak

(*Quercus hemisphearica*), red bay (*Persia borbonia*), cabbage palm, magnolia (*Magnolia grandifloria*), and southern red cedar (*Juniperus silicicola*). Several pines, the loblolly (*Pinus taeda*) and longleaf (*P. palustris*) contribute to the canopy in some areas as well. The most common understory shrubs include the yaupon and wax myrtle, as well as the buckthorn (*Bumelia tenax*).

In the vicinity of Cape Canaveral, a diversity of tropical trees and shrubs begin to increasingly mix with and replace those more adapted to temperate northern areas. However, climate is not the only factor affecting plant distribution on the backdune; soil type must also be considered. Along much of the southeast coast, for example the canopy is formed by a cabbage palm—live oak assemblage, particularly where the soil is composed of moist sand. The temperate understory shrubs common along the northeast coast mix with a diverse tropical flora from Palm Beach County on to the south, but between Cape Canaveral and Palm Beach County lies a transitional region in which the tropical and temperate components mix.

Along the southwest coast, live oak becomes rare and cabbage palm dominates the backdune, along with grassland in low lying areas or a tropical forest on higher ground. Where tropical understory is found, it is generally less diverse than that of the east coast, and sometimes is replaced by a desert-like understory of agave and cacti. Along the northern Gulf coast, the backdune lacks tall forests, and is covered instead by smaller "scrub" oaks (*Quercus myrtifolia* and *Q. geminata*), as well as slash pine (*Pinus elliottii*) and rosemary (*Ceratiola ericoides*). Two other shrubs that are widely distributed along the northern Gulf coast, a mint (*Conradina canescens*) and the woody goldenrod (*Chrysoma paucifloculosa*), also form the backdune plant community found here.

ANIMALS OF THE DUNE

In addition to the numerous insects, spiders and other small invertebrate inhabitants of the soil, a number of larger vertebrate animals make their homes among the dune vegetation, or use this as a key habitat during part of their lives. Shorebirds and seabirds nest in the dunes, and a variety of small rodents and other furry four-legged creatures build their homes and raise their young in this sheltered place. Reptiles—snakes and lizards—also live and hunt here. Most of these are not readily observed—they are small, quick and wary, and are adept at taking full advantage of the cover offered by dune plants. Many shelter in burrows or other secure hideouts during the day, emerging only at night to gather food or conduct other business. The average beachcomber is much more likely to hear their rustlings or see their tracks than to directly observe these ubiquitous but well-concealed creatures of the dune.

Several shorebirds and seabirds nest in Florida's dunes. These include the willet and Wilson's plover (Plate 33c and e), birds that nest in somewhat open areas of sparse grass or clumps of bushy herbs. Other species, such as the Caspian tern (Plate 31), prefer areas of thicker vegetation—usually grasses or small bushes— that provide greater isolation between nearby nests. These and other shorebirds and seabirds are described in Chapter 5.

Among the larger animals often found in Florida's dunes are foxes, raccoons, skunks and opossum, creatures also found well inland. These are nocturnal animals, and the greatest likelihood of seeing them out and about is while driving at night along coastal roads adjoining or bisecting the dunes, particularly in relatively undeveloped areas. They forage and scavenge along dunes as well as nearby beaches. Raccoons in particular are quite accomplished at locating and digging up turtle eggs (Figure 4–16). In some of the most vulnerable areas, sea turtle nests are fenced off to protect them from such predators.

Larger mammals, like raccoons, are highly mobile, and individuals from the mainland and nearby barrier islands interbreed frequently enough to prevent major genetic differences—the first step of speciation—from occurring. However, in some of the smaller and less mobile dune residents this is not always the case, and several types of rats and mice common to Florida's mainland have given rise to distinct subspecies endemic to isolated barrier islands. One of these is the field mouse (*Peromyscus polionotus*),

4-16. *Dune-dwelling raccoons are a hazard to sea turtle nests.*

with five island subspecies now recognized. These "beach mice" are lighter in color than mainland mice, and range from the fore-dune into the scrub, feeding on the fruits and seeds of grasses, herbs and other dune vegetation. Two rats—the cotton (*Sigmodon hispidus*) and rice (*Oryzomus palustris*)—also have developed barrier island subspecies. Thus, the term "beach rat" has a literal as well as figurative meaning! Unfortunately, due to extensive loss of habitat from development of beach property, most of the rats and mice of Florida's barrier islands are now officially listed as threatened or endangered.

5

Life in the Shifting Sand

T he beach is the most transitional of the four zones forming the coastal environment, being an area that is alternately above and below the sea surface. It is also the primary natural habitat of the beachcomber. The general structure of sandy beaches, and the forces that shape them, were discussed in Chapter 2. Here, we will take a look at the living creatures (and their traces) that inhabit Florida's beaches, with emphasis upon those that are most commonly observed.

THE BEACH AS A HOME

As fond as we humans are of beaches, we need to remember that we spend only a small portion of our time there. For creatures that call the place home, the beach is in many ways a highly demanding and unusually inhospitable environment. There is relatively little variation in habitat composition or structure compared to that found in most familiar areas of dry land. To a large degree, this is because there is no vegetation cover to provide food and refuge. Additionally, the substrate itself is generally composed of a relatively homogeneous mixture of fine particles, shaped by sea and wind into a rather uniform low-relief platform that often stretches as an unbroken strip for many miles. Beaches are typically exposed to sustained winds that tend to draw the moisture from living things, intense ultraviolet radiation from which there is little protection, periodic tidal inundation of seawater interspersed with freshwater flushing from rains, and occasional violent pummelling by large destructive storms.

Another major point to consider is that beaches are highly transient places. The very substrate comes and goes with the seasons and is therefore a less-than-ideal place to build a permanent home, even if one seeks protection well below the sand surface. Because all parts of a beach are vulnerable to unpredictable and sudden inundation along with the predictable annual tidal changes, any animal living on or in the beach needs to be able to burrow quickly and/or swim to survive, and must be capable of withstanding exposure to the atmosphere and to the sea at just about any moment. These are not common traits among animals.

With no resident plants and relatively few animal residents, what is there for animals to eat on a sandy beach? The answer is simple—whatever the sea delivers, or each other. As the tide advances and retreats each day, it leaves behind lines of washed-up debris that may have been carried just a few yards or a few thousand miles (Figure 5-1). These lines of debris—called wrack—are composed of the remains of animals and plants and, increasingly, human trash and garbage.

Several parallel lines of wrack may often be seen at different distances from the base of the dune, representing the highest advances of previous tides that carried rich loads of debris. Some of the highest tides may push wrack far enough ashore that it remains untouched by months of successive tides, where it may contribute to the generation of dunes (Figure 4-3). Wrack is a critical resource for many beach animals, providing a source of food, shelter, and moisture, and is often the birthplace of the sandy dunes that form the shoreward border of the beach proper.

All in all then, while it's possible for animals to "make a living" on beaches, it isn't easy. Even so, there are a number of creatures that have adapted well enough to beach environments to thrive there, a tribute to the plasticity and tenacity of life on Earth.

5-1. *The retreating tide leaves behind a line of foam and an unseen trail of minute organic debris—food for tiny beach critters.*

Still, the unusual conditions found on beaches have required adaptations and life-styles particularly suited to such a demanding place, and few successful beach creatures are at home in any other type of habitat. We will now look at the true residents of Florida's sandy beaches.

ABOVE THE DAILY TIDE: CREATURES OF THE BACKSHORE

The beach proper includes that area lying between the highest and lowest *annual* tides normally experienced at that particular location (see Chapter 2). However, the *daily* tidal range is far less than the annual, and thus on any given day, a beach contains two distinctly different kinds of places inhabited by quite different kinds of residents; a dry upper beach or *backshore,* and an alternately submerged and exposed *foreshore,* or intertidal region. During the course of each year, the relative widths of foreshore and backshore often change dramatically.

The dry backshore often takes the form of a distinctive flattened terrace (berm), and is the least hospitable living place in the entire coastal zone (Figure 5-2). At the surface, there is neither rooted vegetation nor the sea to hide behind—little or no shelter from either the elements or the searching eyes of hungry predators. Not surprisingly, few animals rest here, and even fewer spend the entire day. There are a few notable exceptions, however—creatures that have learned to take full advantage of the meager opportunities offered by the backshore.

One beach animal commonly observed amid the dry sands of the backshore is the ubiquitous ghost crab (*Ocypode quadrata,*

5-2. The dry sands of the berm (left) are a demanding environment for living things. Note the distinct scarp (center) separating the flat berm from the sloping foreshore.

Figure 5-3). These comical creatures may be seen peeking from their burrows or scurrying quickly between burrows and foraging sites. Their tiny black round eyes are set on long stalks like little twin periscopes, allowing their bodies to remain concealed in slight depressions in the sand even while they scan the nearby landscape in all directions. They seem more sensitive to motion—a tell tale sign of approaching danger—than to the particular shape or identity of the object moving.

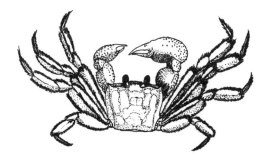

5-3. Ghost crab
(Ocypode
quadrata, x 1).

Ghost crabs rely on quickness as well as camouflage to avoid enemies, and are capable of surprising bursts of speed. They forage heavily on small clams and mole crabs, as well as dead animal remains left upon the beach. Ghost crabs also are capable of feeding on *detritus*—tiny bits of biological debris slowly decaying amid surrounding sediments. Most feeding and foraging occurs during the coolness of mornings and evenings, but ghost crabs will also be seen occasionally leaving their burrows to gather food during midday.

As many a half-asleep sun-worshipper has observed, ghost crabs also work tirelessly to excavate and maintain their burrows, which may extend to over four feet into the sand. Using the claws as shovels and bulldozers, the crab disappears for a bit and then suddenly tosses a claw-full of sand from the open burrow entrance. When too much sand piles up close to the entrance and starts to flow back into the burrow, the crab comes out and bulldozes the sand away, flattening and spreading it about. The whole process makes for an entertaining display on a lazy summer afternoon.

Periodically, ghost crabs must return to the water to moisten their gill tissues. The frequency of such behavior depends to some extent upon the weather—dry hot conditions bring on more rapid water loss. Under normal conditions, the ghost crab may be able to go about its business for up to two days without replenishing gill moisture. Ghost crabs become inactive during the cold winter months, when they remain deep within tightly packed burrows.

The so-called beach hoppers or beach fleas (Figure 5-4) are another group of common residents of the upper beach. These tiny animals are often mistaken for insects, but are actually members of a group of crustaceans called *amphipods*. They are generally found amid clumps of decaying seaweed left ashore as lines of wrack. Although these animals have left the sea, they nonetheless still rely upon gills for respiration, and the gill tissues must be kept moist to function properly. The damp rotting seaweed supplies that moisture, as well as a food supply for beach fleas. When threatened or disturbed, beach fleas respond with a sudden "hop" that carries them a distance of many body lengths.

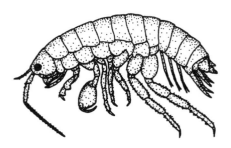

5-4. Beach flea (Orchestia spp., x 7).

THE FORESHORE: A WORLD AWASH

The foreshore—that portion of the beach that is covered and uncovered daily by the changing tide—is quite a different world than the dry backshore (Figure 5-5). Here the sand and its occupants are regularly bathed in nutrient-rich seawater, providing life-giving food and moisture to the odd lot of creatures choosing to call this place home. Much of the life here is microscopic, and well-hidden between tiny grains of sand. These life forms are col-

5-5. The wet sands of the foreshore contain abundant life just beneath the surface.

lectively referred to as the *meiofauna,* an assemblage of organisms far too small to be seen by the unaided eye of the beachcomber, but which nonetheless play a key role in the overall ecology of this zone.

As the tide moves to and fro, so do many of the occupants of the lower beach, animals that prefer to maintain themselves within reach of the swash. Although in many ways more benign than the desert-like backshore, the lower beach nonetheless is also a harsh environment by most standards, and there are relatively few creatures truly at home here. The few animals to be regularly found in the wet intertidal sands of the foreshore are beautifully adapted to thrive there and nowhere else.

Perhaps the most characteristic and ubiquitous dweller of the foreshore is an active little crustacean known as the mole crab (*Emerita* spp., Figure 5-6). Mole crabs are found in this same habitat along most of the east and west coasts of North America and throughout the islands of the Caribbean. The slick-smooth shell, stubby legs, and special digging appendages permit this ani-

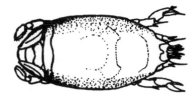

5-6. *Mole crab*
(Emerita spp., *x 1*)

mal to burrow tail-end first through sand with ease. The mole crab feeds on suspended organic particle carried in the swash, trapping and manipulating this food source with a pair of specialized antennae bearing many fine bristles. On occasion, mole crabs are dislodged entirely from the sand and swept up into the water column by strong surge. This is not an insurmountable problem for these hardy little creatures, however, because they are adept swimmers and are able to quickly regain the seafloor.

Most of the mole crabs likely to be seen by beachcombers are females, which grow a good deal larger than the males. Often, they may be seen stranded for a moment at the edge of a retreating wave. They may also be found by digging just below the surface of the wet sand. During spring, females brood masses of orange-coral colored eggs on their abdomens. After several weeks of incubation, the newly-hatched embryos emerge and swim free into the plankton where they undergo further development as they drift. Eventually, the survivors return to the shore to colonize other beaches. Mole crabs are a major food source of several common fishes of the surf zone, and thus are a favored bait for surf fishing.

Another very common and abundant inhabitant of the foreshore sands is the coquina or bean clam (*Donax* spp., Figure 5-7). Although less mobile than the mole crab, these small animals are

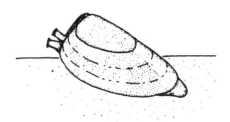

5-7. *Coquina clam*
(Donax spp., *x 2*).

much more agile than one might expect of a clam, and manage to move with the tide so as to stay in a favored position within the swash zone. The coquina clam is equipped with a powerful digging foot and a smooth shell, enabling the animal to glide easily through the wet sand. Coquina feed on plankton and organic debris suspended in the passing water. The water is drawn in through short siphons that extend just above the sand; thus the animal must maintain itself very close to the surface. In turn, coquina clams are preyed upon heavily by fishes and some predatory snails that also inhabit the foreshore.

The most common predatory gastropods of this part of the beach are the olive, moon, and auger snails. Two similar species of olive snails are common in Florida, with the West Indian netted olive (*Oliva reticularis*) dominant in the southern part of the state, and its congener, the lettered olive (*Oliva sayana*, Figure 5-8) replacing it to the north in the Carolinian Province. Both species propel themselves through the moist sand just below the surface as they hunt for coquina, leaving a telltale furrow on the surface that is clearly visible to the trained eye.

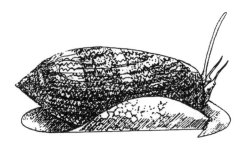

*5-8. Lettered olive (*Oliva sayana*), a predacious snail of the intertidal sands (x 1.5).*

The moon snail (*Polinices duplicatus*, Figure 5-9) hunts in much the same manner as the olive, but feeds by drilling a hole through the shell of its clam prey. The holes are a calling card of sorts, and may be frequently seen in the shells of a number of small intertidal bivalves, including most notably *Donax*. The auger snail (*Terebra* spp., Figure 5-10) sports a distinctive shell that looks something like a drill bit, and occurs along the foreshore as well as in the deeper waters of the surf zone and beyond.

5-9. *Moon snail*
*(*Polinices duplicatus,
x 0.7).

5-10. *Augur shell,* (Hastula
salleana, *x 2).*

This animal also preys heavily upon coquina and other small animals it encounters.

With all these predators about in the surf zone, there is left an abundance of shells whose original owners became the meals of other creatures. There is little waste in nature, and recently vacated shells often become occupied by new owners—the familiar and often comical hermit crabs (Figure 5-11; Plate 25). Hermit crabs occupy a variety of habitats other than the foreshore, including the sands of the dunes and waters of the offshore zone. There are many species, each adapted to take advantage of the opportunities presented in one or another of the various habitats found in the coastal zone.

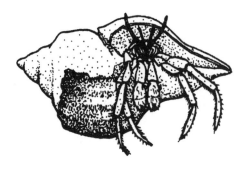

5-11. *The hermit crab*
carries its home upon
its back.

FLOTSAM AND JETSAM: GIFTS FROM THE SEA

The living things we have discussed so far in this chapter are true beach residents, adapted specifically to live out all or most of their adult lives on sandy shores. There is another group of organisms quite likely to be seen on Florida's sand beaches that are not at all in the same category, however. These are creatures adapted to much different environments, but inadvertently carried up onto the beach and stranded there by waves. Some are *bona fide* residents of the open sea, while some are hitchhikers on the remains of other larger organisms, or on man-made debris floating in the open sea. Others are residents of the local offshore zone that have been torn loose from the sea floor or died, and then are washed ashore. Although these animals and plants are native to other ecosystems, we will discuss some of them in the context of beaches because that is where the beachcomber is most likely to see them, and because in total these castaways unquestionably affect the ecology of the beach through their contribution to the nutrients and food supply made available to beach residents.

Taking the time to examine flotsam and jetsam is one of the most rewarding aspects of beachcombing, because it presents an opportunity to discover something *really* odd and unexpected— just about anything that lives in the sea has a chance of ending up on a beach. Because the beach is not the "proper" habitat for these creatures, those that reach the shore alive will usually quickly perish and dry out. For obvious reasons then, the best opportunities for investigating these mysteries of the sea is immediately following storms or periods of sustained strong onshore winds.

Hitchhikers

The upper waters of the open sea may seem a desolate expanse to the naked eye of an observer watching from the deck of a ship. Certainly, there is the occasional fish or porpoise or sea bird, but these are relatively few and far between in most places. For the most part, the remainder of the ocean appears empty and lifeless. But that is largely an illusion. Those of us born to the land are accustomed to seeing and thinking of life in its larger or *macroscopic* forms. But every cup of seawater is alive with many thousands of microscopic creatures that are individually invisible to the naked eye. These include animals, plants, and other simpler organisms such as bacteria and fungi.

Among these microscopic bits of life occur a variety of larval forms of marine animals (Figure 5-12). The production of such pelagic larvae is extremely common among marine animals, because it offers a means of widely dispersing offspring, as well as other advantages. Pelagic larvae are the rule for most marine fishes, as well as many bottom-dwelling invertebrate animals such as corals, lobsters, sea urchins and a host of others. Such larvae are specialized for this pelagic life-style, and bear little if any resemblance to their eventual adult forms. They spend the first part of their lives passively drifting with the currents until encountering proper conditions, an event that triggers drastic changes in form and function as the creatures begin to mature into adult forms.

5-12. *A pelagic developmental stage of a crab, called a megalops larva (x 10).*

For some, the mechanism that triggers the process of *metamorphosis* (change from larval form into intermediate or adult stage) is the encountering of a solid substrate—a piece of driftwood, a floating beer bottle, a clump of seaweed—just about anything that may be clung to. Some of these changelings will only survive a short time without reaching shallow water and gaining access to the seafloor. Others may successfully complete the entire adult portion of the life cycle as drifting hitchhikers. Thus, when we find a piece of wood that has been at sea a long time, it will contain a mixture of "permanent" and transitional life forms. Here, we will confine our discussion to habitual long-term residents of drifting objects—those creatures that will be most commonly encountered and most readily observed.

First and foremost among the animal hitchhikers of pelagic debris are the ever-present barnacles. Two main types are common. The acorn barnacles (*Balanus* spp., Figure 5-13, and *Chthamalus* spp.) have shells consisting of a hard base and six overlapping plates (Figure 5-14) and are fairly low in profile. These animals use specially modified legs to filter tiny planktonic animals from the sea. The legs may be extended or retracted

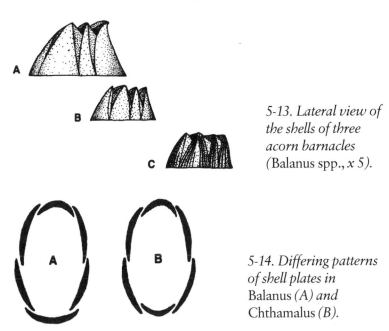

5-13. Lateral view of the shells of three acorn barnacles (Balanus spp., x 5).

5-14. Differing patterns of shell plates in Balanus *(A) and* Chthamalus *(B).*

behind a set of valves that together form a structure called the *operculum,* a device that serves to alternately close off or expose the barnacle's body to the elements. When underwater, the operculum is generally open to allow the barnacle to feed, but when exposed to the air, the operculum seals off the body and prevents it from drying out. It is this ability to withstand periods of prolonged exposure to both sky and sea that makes the barnacle such

an adept hitchhiker on floating debris, as well as a common intertidal inhabitant of solid shores (next chapter).

The stalked barnacles (*Lepas* spp., Figure 5-15) are also common riders of flotsam and jetsam. These differ from the acorn barnacle in the structure of the shell, and in the fact that the animal
attaches to the substrate by means of a flexible stalk rather than a
solid shell base. Despite these structural differences between
stalked and acorn barnacles, however, the basic life-styles and
methods of feeding are much the same.

5-15. *Stalked barnacles,*
(Lepas spp.*) attached to*
driftwood.

Often, one may see fuzzy branched structures growing in a
variety of colors from drifting wood or other debris. Easily mistaken for plants, these are actually hydroid colonies—members of
a group of primitive invertebrate animals related to jellyfish and
corals (Figure 5-16). Like jellyfish, hydroids often bear stinging

5-16. *The hydroid* Obelia dichotoma,
with magnification of colony members
of several types (x 7).

cells used for defense and for the capture of food. Although tiny, some species are capable of producing enough of a sting to warn beachcombers to "look but don't touch." Other forms of sea life likely to form an encrusting layer upon drifting objects long at sea include multi-hued colonies of bryozoans and seaweeds (Plate 9). The latter are often red coralline algae, plants with a crusty skeleton giving them a brittle chalky feel. Alternately, leafy green algae are also common riders of driftwood.

Hitchhikers occur below as well as upon the surface of driftwood, and by breaking apart their rafts, you may observe traces of the handiwork of boring animals, if not the culprits themselves. The burrows and tunnels that perforate the wood are often the work of small boring clams and isopods.

Wash-ups from the Local Offshore Zone

The seafloor of the surf zone and offshore zones are home to a multitude of attached and burrowing creatures that may be torn from their homes and cast ashore by waves and storms. Because the fauna of the seafloor changes substantially between different geographic regions of the state, it follows that the types of creatures likely to be seen on Florida's beaches via this mechanism also may vary greatly from one beach to another. Here we will discuss some of the most common "wash-ups" likely to be encountered.

A group of animals common to much of Florida's coastline are the sand dollars, of which there are several species. These echinoderms appear to be covered with brownish or purplish fuzz when alive, and spend their time just below the surface of the sand in shallow water feeding on organic matter trapped among the sediments. The dried out shells (properly called *tests*) of these creatures are whitish in appearance and bear a distinctive pattern on their upper surface (Figure 5-17). Sand dollar tests are frequently cast ashore, but they are somewhat fragile and it is more likely for beachcombers to find fragments than intact spec-

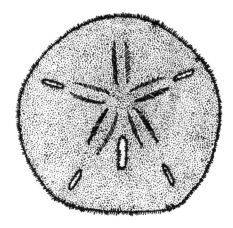

*5-17. The sand dollar
displays a distinctive
pattern on the upper face
of its test (shell).*

imens. If you do find a "keeper," handle it gingerly to ensure its
safe arrival at your home.

Another group of animals you may encounter washed up on
Florida's beaches are known as the octocorals (Plate 26). These
are similar in many ways to the more famous reef-building corals,
but instead secrete a flexible skeleton. The most common octocorals found by Florida beachcombers are the sea fans (*Gorgonia*
spp., Plate 27). These are usually greyish to purple in hue and
have a tough skeleton that will remain relatively intact after drying. For this reason, they are not infrequently used as decorative
pieces. However, be aware that sea fans and all other corals are
protected by Florida law, and you may not legally be in posses-
sion of these creatures, dead or alive, unless you have proof that
they did not come from Florida waters.

Sea stars (Figure 3-13), more commonly called starfish, inhabit sandy and rocky bottoms alike, where they hunt for molluscan
prey. These animals come in a wide variety of sizes and colors,
and may remain relatively intact for long periods after being cast
ashore and dried in the sun.

A truly enormous variety of sea shells may be found cast up on
Florida's beaches. A description of these is well beyond the scope
of this book, and indeed entire volumes have been written on this
subject alone. The reader interested in shell collecting is referred
to the bibliography at the end of this guide.

Wash-ups from Other Shores

A final class of objects often found amidst the wrack of Florida's beaches are the specialized floating seeds of tropical coastal plants, voyagers that may have drifted over hundreds or even thousands of miles before reaching our shores. Most prominent among this group is the familiar edible seed of the coconut palm *Cocos nucifera* (Plate 28), a plant that has in many ways come to symbolize sandy beaches of south Florida and the wider Caribbean. Interestingly, this icon of tropical shores is not native to the region, but was introduced long ago from the far away south Pacific. A group of darkly colored clam-shaped seeds called seabeans (*Mucuna* spp.) are also common wash-ups along our shores. These range from about the size of a quarter to several inches in diameter.

These floating seeds are particularly adapted to sea-borne dispersal, and have consequently developed tough, water resistant coats to protect their living cargos from pounding, dehydration, and saltwater intrusion. They also have internal air pockets to keep them afloat. Floating seeds found on Florida's beaches often originate along the continental mainlands of South and Central America, or in the Caribbean isles. Some are carried beyond the Eastern Seaboard of the United States by the Gulf Stream, and eventually come ashore in the British Isles. They are hardy enough to remain viable throughout these long voyages, and will germinate even after a year at sea provided suitable conditions are encountered.

Wash-ups from the Open Sea

Among the creatures often cast ashore upon Florida's beaches are some delicate forms of sea life adapted specifically to drift on their own about the open sea. Because these fragile creatures tend to quickly dry out and deteriorate beyond recognition once on dry land, we will briefly mention here the most common, and defer a more detailed discussion of them a later chapter, which discusses these wonderful animals and plants in their intact living state. The

categories of such life that the Florida beachcomber will most frequently observe washed ashore are jellyfish, siphonophores (Portuguese man-of-war and by-the-wind sailor), and a pelagic brown
seaweed called *Sargassum.* These are discussed and illustrated in
Chapter 7.

COASTAL BIRDS

Birds are often excluded from popular books dealing with
marine life, perhaps because these animals are almost always seen
in the air or on land, places that are not really part of the ocean.
Nonetheless, many species of birds depend completely or partially upon the sea for food, and play a role in the ecology of marine
and coastal ecosystems through their interactions with other sea
creatures, as well as through their roles in the endless recycling of
nutrients in the sea. Thus, some birds may indeed be considered
sea creatures in a very real sense, and many of the most common
types found throughout the world may be readily observed from
the sandy beaches of the Sunshine State.

A great variety of birds may be seen along Florida's shores.
Birds are unique among animals in their ability to quickly trans-
verse large geographic areas, and thus frequently use a variety of
seemingly remote and disconnected ecosystems for different purposes in the course of a year, or during the life cycle. Many
species observed along Florida's coast are simply transients,
pausing briefly to feed or rest while en route from far away places
like the arctic tundra, Canadian forests, and western prairies of the
U.S. to warmer tropical shores of South and Central America.

In addition to such transients, the Florida beachcomber may
occasionally observe any of a variety of true pelagic seabirds.
These are species that spend most of their lives at sea, often far
from land, and come ashore only to nest. Included in this group
are shearwaters, petrels, storm-petrels, a few species of gulls and
terns, boobies and gannets, and several other groups. These transients and temporary visitors will not be considered here, and
those readers with a particular interest in birds are referred to the

appropriate references provided in the bibliography at the end of this guide for a thorough treatment of the group.

Because coastal resources are diverse, there are many different kinds of places in which to shelter and forage, and many types of food available in each of those places. Consequently, there exists a relatively large number of different groups and species of coastal birds. Additionally, a multitude of wide-ranging species common to inland areas may be seen as well along the coast. The birds of particular interest to Florida beachcombers are those primarily and regularly associated with coastal habitats of the Sunshine State, and we will therefore restrict our discussion to a relatively few such representative species.

Coastal birds may be divided into several groups, based upon the common method of hunting and capturing food. The first of these categories includes those species that float or swim upon the water's surface, and chase down their fish prey underwater with swift and agile swimming ability—the so-called diving birds. The double-crested cormorant (*Phalocrocorax auritus,* Plate 29) is the best example of this hunting strategy in Florida waters. This bird is very common, particularly in south Florida, in bays and estuaries as well as along the beach. It breeds in colonies within mangrove areas along the shores of the mainland and the islands of the Keys. It is an excellent swimmer, and successfully chases down fish from the surface to depths of at least 25 feet. This bird spends a great deal of time sunning and preening itself, either while floating upon the water's surface or perched upon just about any convenient waterside resting place, including beaches, jetties, piers and moored boats (Plate 30).

Aerial searchers represent a second major feeding type of coastal birds. These fly over the coast and sea surface in search of prey, which is then either plucked from the surface while the hunter remains in flight, or chased down underwater by the diving predator. Some species rely primarily on one or the other of these techniques, while others seem equally adept at both.

The gulls and terns, common sights throughout Florida's coastal areas, are birds employing such a strategy with great success. Gulls are larger birds with squared or rounded tails, whereas the smaller sleeker terns usually have forked tails. Common gulls along Florida coasts include the laughing (*Larus atricilla*), ring-billed (*L. delawarensis*), and herring (*L. argentatus*) (Plate 29). The most frequent terns are the least (*Sterna antillarum*), royal (*S. maxima*), sandwich (*S. sandvicensis*), common (*S. hirundo*), and Caspian (*S. caspia*) (Plate 31).

The gulls in particular are a prime example of wildlife whose feeding habit have been radically altered by human populations. In addition to their natural feeding grounds of the coastal zone, garbage disposal facilities throughout the state are now used as a major food source by huge numbers of these birds. Gulls are also attracted to populated beaches where they scavenge leftovers from beach visitors.

Other notable aerial searchers include the brown pelican (*Pelicanus occidentalis,* Plates 29 and 32), the black skimmer (*Rynchops niger,* Plate 33), and the frigatebird (*Fregata magnificens,* Plate 29). The brown pelican is a familiar sight along Florida's coast, and is commonly seen perched atop pilings, wharves, jetties, and the like. Formations of these birds are also often observed flying just above the beach and surf zones (Figure 5-18).

5-18. A formation of pelicans soars above the beach.

Pelicans spot their victims from above, and then fold their wings and plummet straight down into the water to scoop up the prey with unerring accuracy. Occasionally, they will attempt to make a meal out of a fish being used as bait by people fishing from piers, causing the birds to be caught and sometimes injured.

The black skimmer (Plate 33) may be seen throughout the year in Florida, but is most common during the winter months. It spends much of the day simply lounging around beaches, causeways, and waterfront parking lots, and confines the bulk of its food gathering activities to the hours of dawn, dusk, and darkness. It flies very close to the water's surface with the lower half of its bill submerged. Upon striking prey, the top half of the bill snaps shut to complete the capture. This method of hunting is probably most productive at night, when many sea creatures rise from the depths to feed near the ocean surface. The strategy of the black skimmer works best in relatively calm waters, and it frequently forages in protected bays and estuaries.

The frigatebird (Plate 29), also called the man-o'-war bird, is characterized by its unmistakable and distinctive dark angular silhouette. These are among the best gliders in the world, and may remain aloft for many hours by riding thermals, with a minimal expenditure of muscular energy. Frigatebirds are accomplished pirates as well as hunters, and are known for their habits of stealing food from other seabirds in midair and of robbing their nests of young and eggs.

The final category of coastal birds that we shall consider here are the shorebirds that forage and feed by walking about the foreshore, using their bills to probe beneath the sand. For beachcombers, these are the most "watchable" of coastal birds. They are almost always present and active, busily scrambling about the swash zone (Figure 5-19). Some don't seem to mind occasionally having their lower legs submerged by incoming waves, while others appear to go to some lengths to avoid being touched by the sea. These are sometimes quite comical as they race back and forth just beyond the reach of the advancing and retreating swash. Such behavior is most certainly related to hunting rather than keeping dry feet on the beach, as mole crabs and other tasty sand

5-19. Sand-probing shorebirds hunt among the sands of the foreshore.

dwellers may become momentarily exposed or be given away only briefly in response to the movement of the swash. The most common examples of this guild are the willet (*Catoptrophorus semipalmatus*), sanderling (*Calidris alba*), and several plovers (*Charadrius spp.*) (Plate 33).

Our section on coastal birds would not be complete without mention of another group of birds that stalk about hunting prey buried in coastal sediments. These are the large herons and smaller egrets, commonly grouped as wading birds. Although most of this group is more common along the banks of estuaries and on exposed inland mudflats, these birds are also sometimes seen along the outer coast. One in particular, the great blue heron (*Ardea herodias,* Plate 34) is frequently observed along the shores of south Florida and the Keys, often on docks and jetties. These individuals now appear to prefer scavenging leftovers or accepting handouts to their usual hunting behavior.

6

Solid Shores: Rocks, Piers, and Jetties

THE WORLD OF THE ROCKY INTERTIDAL

A lthough the state of Florida is bounded mainly by sandy beaches and muddy embayments, there are a few areas where solid substrates project from, or form short stretches of the shoreline. While solid shores represent but a minor component of the Florida beachcombing experience in terms of extent, such habitats are occupied by a diversity of life forms not to be found on sandy shores. Hence, they offer a real change of pace to those wishing to expand their knowledge of Florida's seashore life. This chapter discusses the general ecological characteristics of solid shores, and describe their most common and readily observed inhabitants.

Solid and sandy shores have far different physical characteristics and place far different demands upon resident life forms, and it is therefore no surprise that by and large, solid shores are inhabited by a different collection of animals and plants from the remainder of the state's coastline. As with sandy shores, one may distinguish between several ecologically distinctive kinds of living places on shorelines composed of solid substrates (Figure 6-1).

There is a comparatively dry area beyond the reach of the tide, but nonetheless exposed to sea wind and spray. This *supralittoral*

6-1. Intertidal zonation is caused by differences in amount of regular exposure to the sea and atmosphere. The dry rocks in the foreground contrast clearly with the dark band beyond (top center) that has been wet by spray and splash.

zone is home to a few land creatures specially adapted to the somewhat harsh conditions found at the edge of the sea, and a few sea creatures that venture here briefly in search of food. There is also a subtidal or *sublittoral* area that is, under normal circumstances, always fully immersed in seawater. This place is home to fishes, sea mammals, and many invertebrates that also occur well into deeper waters.

Between these two extremes lies a third area that is alternately submerged and exposed as the tides wax and wane. This is the intertidal or *littoral* region, a place inhabited by a marvelously adapted set of life forms that are for the most part confined to this tiny band at the margin of the sea. It is this rocky intertidal region that is of primary interest here, for it offers beachcombers unique opportunity to view a variety of true sea creatures going about their daily business in a natural state.

The intertidal may be more rigorously defined as that area lying between the sea level reached during the highest and lowest predictable tides of the year. The word predictable is used here because strong storms or other unusual events may sometimes cause expected actual tides to vary from those expected on the basis of the relative positions of the earth, sun, and moon (see Chapter 1).

Intertidal regions of solid seashores, wherever they occur and whatever their origin or composition, share several fundamental ecological similarities that nearly invariably lead to their being inhabited by similar and predictable groups of organisms distributed in a widely applicable pattern. The key elements determining the composition and distribution patterns of intertidal marine life of solid shores are the nature of the substrate, amount of wave exposure and the relative durations of immersion in seawater, and exposure to the atmosphere (i.e., tidal effects). Is the substrate wood, loosely consolidated sediment, or solid granite? Do prevailing incoming oceanic swells strike this surface first, or is their force first broken to some extent by previous obstructions? Is this spot uncovered during some low tides, and if so how often and for how long?

It is these factors alone or in combination that dictate to a large extent the kinds of adaptations necessary for survival at any particular place within the intertidal zone, and therefore largely determine the nature of the resident biotic communities found there. The adaptations necessary for survival in this narrow and demanding world are limited to a relatively few specialized groups of organisms; hence, the kinds of animals and plants found in rocky intertidal habitats throughout the world are similar to one another, and different from the biotic assemblages of all other coastal and marine habitats.

Although the littoral or intertidal region is generally considered a broad habitat type, or "zone," in itself, on a finer scale there exist distinctive regions or habitats *within* the intertidal, often themselves also referred to as "zones." Most traditional intertidal zonation schemes are based upon tide levels, as might be expected. However, another widely employed scheme is based instead upon the actual observed distribution patterns of major groups of organisms rather than tide levels *per se*. This latter approach is probably more realistic approach in the context of larger geographic contexts, because distribution patterns more accurately reflect key interactions between tidal and other pivotal factors such as wave exposure. In practical terms, either view is acceptable; all such schemes are somewhat arbitrary in any case.

The rocky intertidal marine life of Florida varies to some degree on a regional basis. Regional differences are largely directly attributable to two good reasons; namely, the nature of the solid substrates available for colonization and local annual sea temperature regimes differ markedly in different parts of the Sunshine State. Along the shores of the Gulf Coast and most of Florida's west coast, the only solid substrates to be found are those constructed by people—piers, jetties, bridges and the like. Artificial structures occur along the east coast also, but these share the coast with some natural formations that form rocky portions of the shore. The northeast coast has its *coquina* deposits (Chapter 3), the central east coast has outcroppings of hard substrate attributable to vast numbers of tube-building worms, and south Florida has some shorelines composed of exposed outcroppings of coralline rock. These differences in substrate types, along with the biogeographic differences discussed in Chapter 3, result in some notable differences in the actual species composition of the animals and plants found within the intertidal zone in these different areas, despite broad similarities in the general types of life forms occurring there.

INTERTIDAL ZONATION AND COMMUNITIES

Within rocky intertidal regions, three somewhat distinct kinds of living places or zones are frequently distinguished, each inhabited by somewhat different characteristic assemblages of life forms (Figure 6-2).

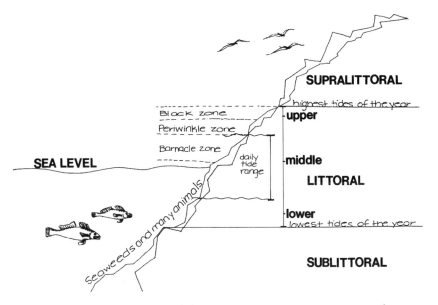

6-2. Representative intertidal zonation pattern common to rocky shores.

The Upper Intertidal

The upper intertidal includes an area regularly wet by sea spray carried by wind or released from breaking waves, as well as a somewhat lower area that becomes immersed during extreme high tides. This zone is sometimes also called the *supralittoral fringe,* because it represents the extreme upper margin of the littoral region. Here, the paucity of moisture or nutrients from the sea combine with prolonged direct exposure to the sun and wind to make for unusually harsh living conditions and the least hospitable of intertidal habitats.

The upper reaches of this zone are generally devoid of sea life, save for the occasional foraging crustacean. But towards its lower margin, the first hesitant traces of marine intertidal life begin to appear. The first sign of life is a darkish, scum-like coating on the rocks that contrasts notably with the bare surfaces just above. Sometimes referred to as black algae, this living film is actually composed of members of a group of single-celled organisms called the *blue-green algae,* one of the most primitive life forms to be found on the planet, and one of the hardiest. Although they are capable of photosynthesis, these are not among the true algae that are part of the plant kingdom. Blue-green algae are so primitive and unique that modern biologists place them in a separate kingdom along with the bacteria, with whom they share certain structural and functional similarities.

Blue-green algae are noted for their ability to survive in places intolerable to all other organisms, such as hot springs and high mountain peaks, so it is not surprising to find their kind in this most demanding of marine habitats as well. Different types display different growth forms. Some grow as individual cells, while others form globular or filamentous colonies. The actual structure of these tiny cells is only apparent under a microscope however, and all the unaided eye of the beachcomber will discern is an unstructured black slimy film that indicates the presence of these humble but amazingly adaptable life forms.

The other main group of marine creatures frequently found inhabiting the lower portion of the upper intertidal are members of a particularly hardy group of marine snails—the periwinkles (*Littorina* spp.). Periwinkles have three or four whirls in their shells, and come in a variety of colors and patterns. Probably the most common member of this group in all parts of Florida is the zebra periwinkle *Littorina lineolata* (Figure 6-3). Periwinkles use a muscular foot protruding from the shell opening to creep about the rocks as they graze on encrusting algae, which is scraped off with a flexible toothed tongue-like structure called a *radula.* Like many

6-3. *Zebra periwinkle (*Littorina lineata, *x 5).*

marine snails, periwinkles also have a protective structure called an *operculum* attached to the foot. When sheltering, the operculum acts as a mechanical barrier as well as a watertight hatch to prevent the animal from drying out. So effective is this mechanism and so hardy are these animals that they have been known to survive well over a month without exposure to seawater.

A final group of animals that might be seen in the upper intertidal are small mobile crustaceans—crabs and their smaller relatives the isopods. These spend most of their time in moist shady underside of the rocks, but occasionally venture forth in search of food. The porcelain crab (*Petrolisthes,* Figure 6-4) is one such animal. The isopods are relatives of the familiar "pill bugs" of the land. The most common in this zone is the so-called "sea roach" *Ligia exotica* (Figure 6-5), an animal frequently mistaken for the terrestrial roach because of its habit of hurriedly avoiding exposure to light, as well as its general appearance. If one is uncovered by turning over a rock, it will quickly scurry about in search of darker recesses. Nonetheless, these are true sea creatures, using gills for respiration.

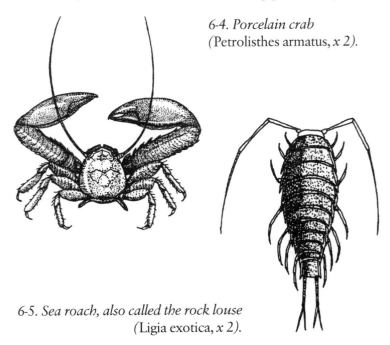

6-4. Porcelain crab
(Petrolisthes armatus, *x 2).*

6-5. Sea roach, also called the rock louse
(Ligia exotica, *x 2).*

The Middle Intertidal

The middle intertidal (also called *mid-littoral*) region includes that area covered and uncovered by most tides. While dehydration is not so much a problem as in the upper intertidal, organisms residing here must still contend with hours of direct exposure to the sun and the atmosphere, as well as with the forces of breaking waves that tend to tear them from the substrate.

There are usually two bands recognizable within this part of the intertidal, each characterized by a different collection of marine organisms coating most of the rock surface. The upper band boasts a carpet of acorn barnacles, described in Chapter 5. In Florida, the most common species found here are the fragile barnacle (*Chthamalus fragilis*), the ivory barnacle (*Balanus eburneus*), and the striped barnacle (*Balanus amphrite,* Figure 5-13). The fragile barnacle is much the smallest, with a diameter of less than ½ centimeter at maturity. Fully grown individuals of the two *Balanus* species grow to at least five times that size, and are easily distinguished from one another by differences in coloration. The ivory barnacle is, as the name suggests, white in color, while the striped barnacle has purplish stripes radiating from the top of its shell plates to their bases.

Below the so-called "barnacle zone" lies an area in which fleshy marine algae, also known as seaweed, dominates the surface of the rocks. Seaweeds are classified into three main groups: browns, reds, and greens. These are true plants, and each group derives its characteristic color from the particular photosynthetic pigments it carries, for each is adapted to somewhat different types of light. In the sea, red light is quickly extinguished, followed by yellow and eventually green and blue, which penetrates farthest in clear tropical seas. The different groups of seaweeds have pigments best suited to the predominant solar wavelengths reaching different depths in the sea. Green algae are best adapted to the shallowest waters, browns to intermediate depths, and red algae are able to use the deep-penetrating blue wavelengths.

These generalizations should not by any means be taken strictly however, for most seaweeds contain a number of photosynthetic pigments in addition to the dominant type, and are therefore able to use sunlight over a broad range of depths. Commonly, intertidal regions are occupied by a combination of all three groups. Green algae become scarce at intermediate depths, and at great depths only the reds remain. Most of the world's larger, more spectacular seaweeds flourish on rocky coasts bathed in cool water of temperate coasts. Tropical and subtropical species found in Florida are generally smaller and far less conspicuous than those blanketing intertidal regions of Maine and California.

Mixed among the dominating barnacles and seaweeds of the middle intertidal may be found a wide variety of other creatures, including oysters, mussels, sponges, crabs, sea anemones, starfish, hydroids, bryozoans and a number of types of marine snails. Hermit crabs (Figure 6-6) are ubiquitous, and maintain themselves close to the waterline. The chiton (Figure 6-7) is a slow-moving herbivorous marine snail with a characteristic shell composed of eight serially arranged plates. These may be found tightly clinging to the rocks throughout most of the middle intertidal along solid shores of the eastern and southern parts of the state.

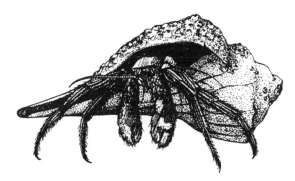

6-6. An intertidal hermit crab in its barnacle-encrusted shell.

6-7. The chiton is an unmistakable dweller of the intertidal.

The precise composition of the diverse community of the middle intertidal varies substantially with location, with overall diversity generally increasing towards the tropics. Descriptions of the many species that may be seen in this zone is well beyond the scope of this volume, and readers wishing to become proficient at identifying particular species occurring here are referred to the references listed in the bibliography at the end of this guide.

The Lower Intertidal

This region, also called the *sublittoral fringe,* extends from the lower margin of the middle intertidal down to the level exposed by the lowest tides of the year. This is a fully submerged habitat for the vast majority of the time. Organisms living here are exposed to constant wave action and the searching eyes and mouths of hungry fish, but their time of exposure to the atmosphere is limited to relatively rare and brief periods.

In the northern part of the state, this region is inhabited by representatives of most of the same groups of marine life found in the middle intertidal, along with an additional diversity of subtidal invertebrates and fish. In south Florida, the lower intertidal explodes in a profusion of tropical animals and plants belonging to many groups, including reef-building corals and octocorals (Plate 5). In areas where the rock surface forms a more or less vertical wall, encrusting calcerous red algae often form a so-called "pink zone" extending to the lowest limits of this area. Here, tube-building worms and snails also proliferate, along with sea anemones, chitons, urchins, crabs, and all three major types of seaweeds.

Although the rocky intertidal region is unquestionably ruled by the invertebrate segment of the animal kingdom, it is in the lower intertidal that the first representatives of the vertebrates begin to assert themselves. Just below the surface of the sea, small marine fishes find living space and food amid the rich algal gardens and invertebrate hordes. The most obvious and ubiquitous fishes living here are the small territorial damselfishes of the Family Pomacentridae (Figure 6-8a). These are usually no more than a few inches in length. Most species have highly similar body shapes, and bear dark hues as adults. As is often the case with reef fishes, juveniles tend to have quite different coloration than their adult counterparts. Damselfish are common in almost all reef and reef-type environments, including rocky shores, rock outcroppings, and jetties.

Two other fish families frequently represented in rocky subtidal areas are the even smaller gobies (Family Gobiidae) and blennies (Family Blennidae) (Figure 6-8). Unlike the active and free swimming damselfishes, these are cryptic in color and habit, and not as readily observed. Many shelter within the cavities of sponges or fashion burrows in other substrates, only occasionally venturing forth (Plate 35). Gobies are primarily carnivores, feeding on small invertebrates, while blennies tend to be omnivores, eating smaller animals and seaweeds.

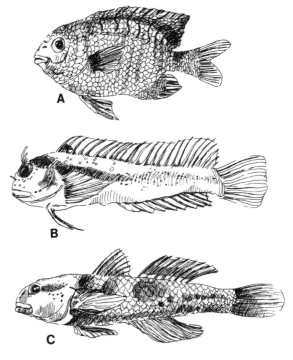

6-8. Common intertidal fishes: (A) damselfish;
(B) blenny; (C) goby.

7

Beyond the Beach: Life at Sea

There are a variety of animals and plants that may occasionally be observed from the beach, but belong to another world entirely. These are the living inhabitants of the open sea, visitors who briefly come near enough to shore to be seen, and then silently move on. Their proximity to the beach is usually temporary and somewhat unpredictable, and they offer the beachcomber a brief look at the ocean world beyond the shore, and a sense of its mystery and magnitude. This chapter is devoted to such life forms.

PELAGIC DRIFTERS

Among the creatures frequently drifting near the shore under the influence of wind and currents are some that spend their lives passively drifting about the surface of the open sea. A variety of jellyfish fall into this category, and these may occur in huge numbers in offshore waters at certain times of the year (most commonly in summer and early fall), and under the right conditions may wash up on nearby beaches *en masse*.

Jellyfish are very simple animals, with most of the body mass composed of a muscular bell used to propel the animal through the water (Plate 4). A ring of tentacles bearing stinging cells trails from the margin of the bell, and surrounds the mouth. Swimming is not used by these animals to traverse horizontal distances, but rather to counteract their tendency to sink below the upper well-lit waters where their food is most abundant.

The most common jellyfish in Florida waters are the cannon-ball (*Stomolophus meleagris*) and the moon (*Aurelia aurita*). The cannonball jellyfish (also called the cabbagehead) grows to about 10 inches across, and is brownish in color and without trailing tentacles (Figure 7-1). The moon jelly has a bluish-tinted bell up to about 15 inches in diameter, and sports a ring of very short tentacles around the bell margin, along with a distinctive four-lobed reproductive structure clearly visible within the transparent bell (Figure 7-2). Fortunately for Florida swimmers, neither of these two species have a dangerous (to humans) sting.

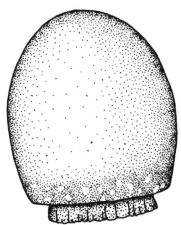

7-1. Cannonball jellyfish, also called cabbagehead (Stomolophus meleagris*).*

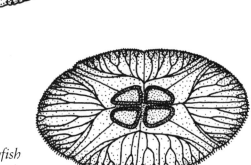

7-2. Moon jellyfish (Aurelia aurita*).*

A much more formidable animal drifting into the surf zone and washing ashore on Florida's beaches from time to time is the unmistakable Portuguese man-of-war (*Physalia physalis,* Figure 7-3). This is one animal well worth learning to recognize and avoid touching, for it delivers one of the most powerful and painful stings of all sea creatures. The Portuguese man-of-war is not a jellyfish, but a related form of colonial hydrozoan known as a siphonophore. It may be easily distinguished by its bright bluish or purplish gas filled float, a structure used to buoy up the colony as it sails the open seas. The floats commonly seen in Florida range in size from about that of a golf ball to that of a coconut.

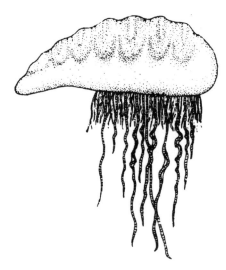

*7-3. Portuguese man-of-war (*Physalia physalis*).*

A painful mistake often made by Florida beach visitors, particularly by small children attracted to the colorful balloon-like float, is to assume that because the colony is ashore and perhaps dead, it is perfectly safe to touch. Nothing could be farther from the truth. The stinging tentacles are often colorless and may trail for 10–20 feet below the float, or be broken off and detached when the colony is washed ashore. Be warned that even when completely separated from the rest of the colony, the tentacles are perfectly capable of discharging their stinging cells.

It is wise to be very careful and alert when swimming along a beach on which these creatures have been recently observed. If stung, one should seek treatment from the nearest lifeguard if possible. In severe cases, treatment by a physician may be advised. As immediate first-aid, a topical pain killer or ammonia is frequently applied directly to the affected area of the victim's skin to reduce pain and irritation. Antihistimines may provide further relief.

Another siphonophore somewhat less frequently seen than *Physalia* is the smaller *Velella velella,* commonly known as the by-the-wind sailor. This creature grows to about three inches or so, and is named for the distinctive float that is modified to look and function as a sail (Figure 7-4).

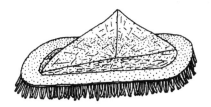

7-4. By-the-wind sailor (Velella vellela*)*.

THE WORLD OF SARGASSUM

A unique and complex entire community of marine life may be frequently observed along Florida's coast, where it has drifted from its main domain—the open waters of the Caribbean Sea. The community is based upon a marine algae known as *Sargassum,* a pelagic seaweed that serves as home to a unique and marvelous assemblage of animals, many of whom are found in this habitat alone (Figure 7-5).

Clumps of *Sargassum* frequently wash ashore along all parts of the Florida coast, where they may be examined high and dry by beachcombers. However, the beach is not really a good place to see the resident animals, who by then have mostly fled or died. Typically, by the time the beachcomber stumbles upon a clump of beached *Sargassum,* there is little more to observe than a notably unspectacular pile of rapidly drying and decaying sea-

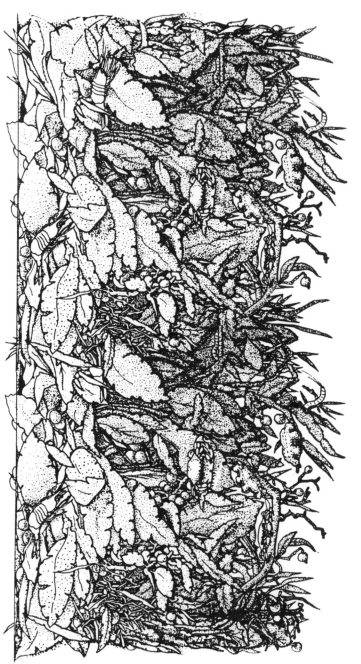

7-5. The living Sargassum community.

weed (Figure 7-6) that does little justice to the unique beauty and complexity of the living community. Fortunately for Florida beachcombers, this world may be easily viewed at leisure while still intact and just a short distance from shore.

7-6. The dehydrated remains of Sargassum *are common along Florida's east coast beaches.*

Sargassum itself is a type of brown seaweed that in life has a yellowish-gold ground color. It has rather small flat and rough-edged blades that serve the same light-collecting function as do the leaves of terrestrial plants, and many small air-filled bladders that give it the buoyancy needed to raft its way across the sea. Careful examination reveals that the plant body is in many places overgrown by a more lightly colored fuzzy coating that feels slightly rough to the touch. These are bryozoans—the moss animals described in Chapter 3. Hydroids also frequently encrust the plant surface.

The best way to examine the *Sargassum* community is with a white bucket, mask and snorkel in about waist-to-chest deep water. The most prolific clumps will be those that have most recently been swept into the surf zone. First, put your mask on and position yourself about a foot away from the seaweed. You will undoubtedly see a number of small fish darting about just below the plant. These will most frequently be filefish (Family Balistidae), some of which match the gold color of the *Sargassum,* as well as young stages of some of the swift pelagic jack family (Carangidae) (Figure 7-7). You needn't worry about fright-

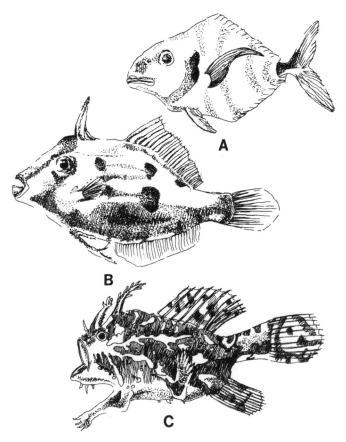

7-7. Fishes of the Sargassum *community: (A) juvenile jack; (B) filefish; (C) sargassumfish.*

ening them away—they have nowhere else to go and will not leave their haven unless there is no other choice.

If you now steady the clump by holding it gently from above and bring it a bit closer, you may begin to see some of the well-camou-flaged invertebrate animals creeping about the plant body. You must look very closely, for these are beautifully colored to match the gold of the plant, and many complete the disguise with irregu-lar, off-color patches that simulate the appearance of encrusting hydroids and bryozoans. Here you may find the sargassum anemone (*Anemonia sargassensis*), the sargassum crab (*Portunis sayi,* Figure 7-8), and a like-colored shrimp (*Latreutes fucorum*).

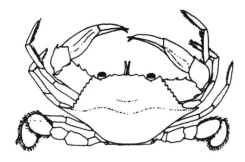

7-8. Sargassum crab
(Portunis sayi, x 2).

Perhaps the greatest prize of all to be found clinging to the lower surfaces of *Sargassum* however is the sargassumfish *Histrio histrio* (Figure 7-7). This is one of the best examples of concealing col-oration and disguise to be found in the animal kingdom. The over-all body color is gold with mottled patches of lighter hues, as with the crabs and shrimp previously described. But the sargassumfish has refined the disguise even further, with its lumpy shape and blade-like fins providing an irregular and most un-fishlike body form. The only symmetrical part of the body that is exposed is the eye, and even that has been disguised by a series of alternating dark and light bands radiating outward in all directions from the pupil.

Needless to say, the sargassumfish relies on stealth and camou-flage to take advantage of unwary prey. It is an able swimmer, but frequently prefers to instead slowly creep about the *Sargassum* using its unusual pectoral fins much like arms as it positions itself to strike another fish or one of the many resident invertebrates. Sargassumfish are tireless and voracious hunters. If several are placed in an aquarium along with their seaweed home, in time they will hunt one another down until only the largest remains.

After you have seen all you are able with your mask, submerge your bucket below the clump, and lift the rim just above the sea surface. If you now gently lift and shake the clump, you will find that your bucket may contain some of the animals described. When you have seen enough, gently place the clump back in the bucket and return the contents to the sea.

MARINE MAMMALS

These amazing sea creatures have received no shortage of attention of late, and most readers will be generally familiar with many of the most common types. Scientists think sea mammals are descended from animals that were land dwellers. Some, such as whales and porpoises, have no remaining dependence upon terrestrial habitats at any time in the life cycle, while others such as sea lions and walruses still return to the land for courtship and the bearing and raising of their young.

Marine mammals are found in virtually all oceans from the equator to the polar seas, and although most are marine, some are found in estuaries and a few even in freshwater. Although there are rare historical records of unusual sightings of a wide variety of sea mammals near Florida, only a relative few are common. The two major groups regularly visiting or living in Florida waters are the whales (Order Cetacea), and the manatees and dugongs (Order Sirenia).

The whales are divided into two main subgroups; the baleen whales and the smaller toothed whales, a group that includes the familiar dolphins and porpoises as well as the now famous killer whale and other less familiar species. While most marine mammals are strictly or primarily carnivorous, one of the more common Florida resident's, the manatee, is a herbivore. Although many species were once ruthlessly hunted, in some cases to the brink of extinction, virtually all marine mammals are now protected under the U.S. Marine Mammal Protection Act, and many are making remarkable comebacks.

Florida waters are not believed to be the home of any of the baleen whales, although sightings of transient individuals are not uncommon. A number of species appear to migrate regularly from the Gulf of Mexico to the open Atlantic through the Florida Straits (located between the Keys and Cuba), and these might be observed while fishing offshore, but rarely if ever seen from the beach unless they become stranded.

Several of the toothed whales are regularly seen off our coast, and Florida may be considered home to populations of these. The most common by far is the bottlenose dolphin (*Tursiops truncatus,* (Plate 36B), a cosmopolitan species occurring around the world in tropical and subtropical seas. The classification of this dolphin is currently undergoing extensive study, and it is possible that there may actually be two species, a smaller and larger version, now included under the same name. Most of the bottlenose dolphins found in Florida are of the smaller type.

The bottlenose dolphin is a darkish grey color on the upper surface shading to a light grey below. This is the dolphin commonly found in marine parks and featured on television and in movies. Those seen in Florida seldom exceed a total length of about 8–9 feet. These animals live in herds of up to 100 animals, but smaller family groups of three to twenty are the most common groups found in state waters.

The dark dorsal fin that protrudes from the water when the animal is at the surface often causes alarm among Florida swimmers, because it is mistaken for the fin of a large shark. Dolphins and sharks may be readily distinguished from a distance with a bit of practice. The shark has a vertically oriented caudal (tail) fin that often breaks the water right behind the dorsal fin, while the caudal fin of the dolphin is oriented in the horizontal plane and will not be mistaken for a smaller second "dorsal" fin as in the shark.

The dolphin will repeatedly arc its way through the water, with both the top of its head and dorsal fin observable in succession, whereas sharks tend to slice through the water at a more constant level, with only the dorsal and caudal fins breaking the surface. This difference is related to the fact that a dolphin is a mammal and must regularly expose the "blowhole" located on the top of its head to exchange air with the atmosphere. Sharks, in contrast, are fish and exchange gasses for respiration directly with the surrounding sea through their gills; thus they have no need to surface. Another telltale clue is that dolphins almost always travel and hunt in groups, whereas sharks are almost always solitary. Although these tips may in time allow you to tell one from the other, until you become proficient it is probably best to practice distinguishing dolphins and sharks from the shore!

The bottlenose dolphin engages in highly complex social interactions within its group, and has a well developed language consisting of both sound and other forms of behavior. The animal is often seen by beachcombers at the outer margin of the breaker zone where it feeds on the small baitfish that are common there. Sometimes, the lucky beachcomber will be fortunate enough to see dolphins apparently "surfing" just for the fun of it, or leaping clear of the sea surface (Figure 7-9).

Although several other species of toothed whales may be occasionally seen passing through Florida waters, the only one that appears to be a regular feature is the short-finned pilot whale (*Globicephala macrorhychus,* Plate 36c), a species that ranges throughout the world's oceans from the tropics to temperate latitudes. It is quite common along the eastern coast of the United States as well as in the Gulf of Mexico, and is frequently involved in mass strandings on these coasts. Fully grown females may reach a maximum length of some 13 feet, while males may attain lengths of 17 or 18 feet. The species is carnivorous, feeding mainly on squid.

The only other sea mammal common to Florida waters is the gentle Florida manatee, an animal that is generally considered a subspecies of the more widely distributed West Indian manatee (*Trichechus manatus,* Plate 36a). The once large population of

7-9. The Atlantic bottlenose dolphin may often be seen playing in the surf zone.

these creatures has declined dramatically coincident with the increasing population of the Florida peninsula. A combination of hunting, habitat loss, pollution, and an alarming increase in the number of serious collisions with power boats have all contributed to the problem. Boat collisions alone are now estimated to account for one third or more of all manatee deaths on an annual basis. There are probably only about 1,000 individuals remaining, and growing concern for the future of this animal has led to its designation as an endangered species. Areas where boaters must use extra caution and slow speeds are conspicuously posted throughout the state.

These strict herbivores spend most of their time grazing peacefully upon seagrasses growing in shallow marine and estuarine coastal waters. They also regularly enter freshwater where they happily devour great quantities of freshwater vegetation. Movement patterns are clearly related to seasonal temperatures in Florida. Although they are year-round residents of the state, they move south during cool fall and winter months and return to the northern parts of the state in the warmth of spring and summer. During particularly cold periods, manatees frequently move into the comparative warmth of coastal freshwater springs, a visit that presents one of the best opportunities for viewing the animals close-up. Manatees are also often seen around the docks and channels of Florida's many coastal marinas.

FISHES

A great variety of fishes occur in waters adjoining the state of Florida. The Florida Keys alone are known to be visited by over 500 different kinds. Clearly, such diversity is well beyond the scope of this book to even begin to approach, and the interested reader is referred to any of the several highly useful fish identification references provided in the bibliography. Here we will confine our discussion of fish life to a brief mention of a few species most likely to be included in the catches of the numerous surf

(text continued on page 143)

Plate 26. Octocorals are commonly mistaken for plants.

Plate 27. A sea fan spreads its branches to collect plankton in an undersea canyon.

Plate 28. The graceful coconut palm (Cocos nucifera).

135

Plate 29. Common shorebirds: (A) brown pelican; (B) laughing gull; (C) herring gull; (D) ring-billed gull; (E) frigatebird; (F) double-crested cormorant.

Plate 30. Cormorants resting on marina rocks in the Florida Keys.

Plate 31. Common terns of Florida: (A) royal; (B) least; (C) caspian; (D) sandwich; (E) common.

Plate 32. Brown pelican (Pelicanus occidentalis).

137

Plate 33. Common
shorebirds:
(A) black skimmer;
(B) sanderling;
(C) willet;
(D) semipalmated plover;
(E) Wilson's plover.

A

B

C

D

E

Plate 34. A great egret stalks
a dock in the Florida Keys.

138

Plate 35. A blenny peers from his coral shelter.

Plate 36. Common marine mammals of Florida: (A) the West Indian manatee; (B) the Atlantic bottlenose dolphin; (C) the short-finned pilot whale.

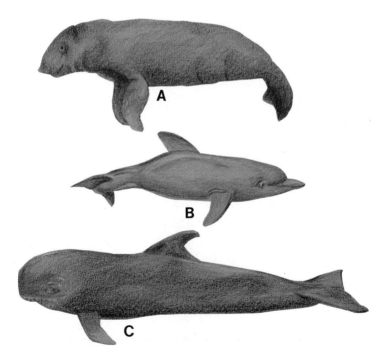

Plate 37. Loggerhead turtle (Caretta caretta).

Plate 38. Green turtle (Chelonia mydas).

Plate 39. Leatherback turtle (Dermochelys coriacea).

Plate 40. Coral polyps extend feeding tentacles at night.

Plate 41. The coral polyp tentacles are withdrawn by day into the protective skeleton.

Plate 42. Close-up view of the surface of a living coral colony. Hundreds of polyps are in view.

Plate 43. A coral reef is made up of a diverse assemblage of plants and animals.

Plate 44. A patch reef surrounded by sand and seagrasses of the lagoon.

Plate 45. *Seagrass and green algae bloom in calm shallow waters of south Florida.*

Plate 46. *A close look at the leaves of the red mangrove* (Rhizophora mangle).

(text continued from page 134)

fishermen that line the Florida coast (Figure 7-10), and therefore observed by beachcombers, albeit in a most unnatural state.

Common inshore fishes of the surf zone are for the most part relatively small, in the order of a foot or so in maximum total length, silvery in color, and adapted to feed upon the most common invertebrate animals of this habitat, such as mole crabs and bean clams. Some are considered good gamefish and excellent eating (Figure 7-11). Among the most sought after of these are the Florida pompano (*Trachinotus carolinus*), the whiting (*Menticirrhus saxitalis*), the red drum or redfish (*Sciaenops ocellatus*) and the bluefish *Pomatomus salatrix,* a darkly colored seasonal visitor to Florida shores that occurs in great numbers when present.

Surf fishing is a popular pastime and a good way to spend a day at the beach. If you decide to try it for yourself, first make sure that you are aware of all applicable laws. Regulations concerning

7-10. Surf fishing is popular along the Florida coast.

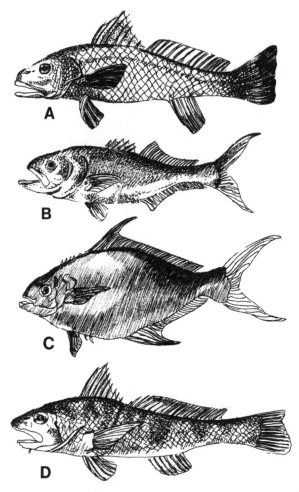

7-11. Common gamefishes of the surf zone: (A) redfish;
(B) bluefish; (C) Florida pompano; (D) whiting.

fishing licenses, gear restrictions, seasons, bag limits, and har-
vestable species are undergoing constant revision. A complete
update and briefing may be readily obtained at any reputable bait
and tackle shop, or from any office of the Florida Marine Patrol.

8

Sea Turtles: Ancient and Endangered Visitors

INTRODUCTION TO SEA TURTLES

Of all the odd and wonderful creatures to grace Florida's beaches, none are more worth waiting for or watching than sea turtles. Visitors from another time and another world, no less than five of the world's seven recognized species of sea turtles might be seen along our coast.

Beachcombers often observe sea turtles either swimming just beyond the waves, or actually making their way up the beach to dig nests in the deep sand and deposit their eggs. Florida contains some of the best turtle-watching beaches in the world, and witnessing the return of these gentle giants should be on the "must" list of every summer beachcomber.

The past several centuries have been hard on sea turtles. Throughout the Caribbean, they have been hunted mercilessly for their meat, shells, and eggs. Populations once numbering in the millions have been decimated by hunting and destruction of critical nesting habitat, until all species are now accorded endangered or rare status.

Florida's beaches represent one of the western hemisphere's most critical nesting sites for the endangered loggerhead turtle, the only nesting site for green turtles to be located in the continental United States, and an occasional nesting site for small numbers of hawksbill and leatherback turtles. The conservation of the world's remaining sea turtle populations ranks high among the priority lists of numerous international conservation organizations, and has many ramifications for coastal development.

Sea turtles are among the largest of living reptiles, ranging in size at maturity from the smallest (ridley) at just under 100 pounds to the largest, the leatherbacks, which may reach a length exceeding eight feet and a weight approaching a ton. They are externally quite similar in physical appearance to common land turtles. Unlike those of their landlocked cousins, however, the heads and flippers of sea turtles are not designed to be fully retractable into the formidable body armor. The shells of sea turtles consists of an upper or *dorsal* portion called the *carapace,* covered in all but the leatherback with a series of hard plates called *scutes.* The ventral or lower section is called the *plastron,* and is joined to the carapace by cartilage.

Because of their common ancestry, all sea turtles share many general similarities. But the group has also diverged over the last 150 million years, and each modern species has developed unique physical, ecological, and behavioral characteristics that serve to distinguish it from the others. Different kinds of sea turtles are quite easily recognizable by a number of distinctive physical traits (Figure 8-1). One of the most useful features in this context is the structure of the carapace. Different species also differ in behavioral traits, such as food and habitat preferences, as well as the timing, location, and other features of their nesting patterns.

The early life histories of sea turtles are not well known. Immediately upon hatching, the baby turtles scamper quickly from the exposed beach and into the waiting sea (Figure 8-2). This almost always takes place at night, when the hatchlings are far less vulnerable to prowling sea birds. Upon reaching the surf, they are by no means out of peril; fishes and sea birds take a regular toll during this exodus to deeper water.

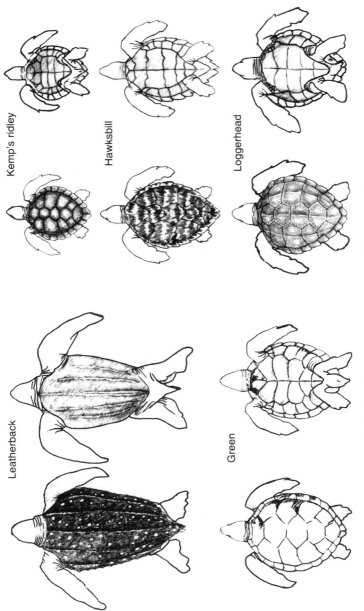

Kemp's ridley

Hawksbill

Loggerhead

Leatherback

Green

8-1. Florida's sea turtles: top and bottom views.

8-2. Newly hatched turtles make their way to the sea.

It is believed that the hatchlings of most species proceed beyond the continental shelf to the deep ocean. There, they spend their early months drifting passively near the surface of the sea, taking advantage of lines of seaweed and other floating debris to gain shelter and food. This life-style—allowing the wind and currents to determine one's destination—is called *pelagic,* and is common among the larval and young stages of many marine animals.

In time, the young turtles re-enter shallower coastal seas, including bays and estuaries, where they begin to use food sources living on or near the sea bottom. Some populations appear to undergo the majority of development into mature adults in a single particular feeding ground, while others are not so regimented and may relocate to a number of different feeding grounds as they grow to adults.

Sea turtles require an unusually long time to reach reproductive maturity. Species and individuals vary of course, but current estimates place the first reproduction at some 15 to 50 years after hatching. When the nesting season arrives, mature adults journey from the feeding ground to a particular nesting beach, often the same one from which they themselves were hatched. In some populations these nesting migrations occur over large distances.

The hundreds of miles of broad sandy beaches that border Florida are ideal nesting sites for some species (Figure 8-3), and the Sunshine State accounts for the most sea turtle nesting that occurs in the continental United States, including about 90% of the loggerhead nests and virtually all of the green turtle nests. The area of greatest nesting concentration occurs along the east central coast, from New Smyrna Beach to Boca Raton. West coast beaches with frequent nesting occur along the south-central coast from Sarasota to Boca Grande, and Cape Sable farther to the south.

FLORIDA'S SEA TURTLES

Of the five species of sea turtles that visit Florida's shores, four are known to nest there; the fifth—Kemp's ridley—has only been observed offshore. Two of the sea turtles that nest on Florida's beaches (leatherback and hawksbill) do so very infrequently, and observations of nesting behavior in these species are unlikely by

8-3. Florida's broad sandy beaches are a perfect nesting site for sea turtles.

any but the most devoted, persistent, and just plain lucky of amateur naturalists. For that reason, this section focuses primarily on the two sea turtles that the average beachcomber stands a reasonably good chance of observing first-hand—the loggerhead and green turtles. Notes on the other three species found in Florida waters are provided for comparative purposes, along with a recognition guide to all five.

Loggerhead Turtles

The most abundant nester on Florida's shores by far is the loggerhead turtle, *Caretta caretta* (Plate 37). Its name is derived from the unusually large skull needed to support the powerful jaw musculature. After the pelagic phase of the life cycle has been completed, loggerheads feed primarily on shelled invertebrates—mollusks and crustaceans—dwelling on and around rocky and reef-like substrates. Adults average around three feet in length and reach a weight of around 300 pounds, but individuals nearly 4 feet in length and weighing around 400 pounds have been observed.

Florida is the second most frequented nesting site in the world for this species, with an estimated annual contingent of some 10,000 to 20,000 females making their way ashore. Nesting occurs mainly along the east central coast from April to September, with peak concentrations achieved in July. The feeding grounds of the loggerhead turtle are widely dispersed throughout the Caribbean, with major sites located as far away as the Dominican Republic. The Gulf of Mexico and the Bahamas also support large numbers of feeding loggerheads.

In the vast majority of cases loggerhead nesting occurs during the hours of darkness. However, daylight nesting behavior has also been observed in Florida, albeit on rare occasions. Typically, females preparing to nest take up waiting positions just beyond the surf zone during late afternoon and evening. When darkness is complete, the turtles swim ashore and literally drag themselves across the beach, propelled by flippers designed for swimming rather than walking. Loggerheads, as do all other sea turtles, leave a distinctive track that allows biologists to census the numbers coming ashore each night on a given beach by simply counting the tracks during the early morning, before the incoming tide and beachgoers have the chance to obliterate them (Figure 8-4).

As they move up the beach, loggerheads appear to periodically "sniff" the sand as though looking for telltale signs that this is indeed the "right" beach, and that environmental conditions appear to be satisfactory for nesting. They are discouraged by bright lights and a lack of dune vegetation, two of the hallmarks of ill-conceived coastal development. At some point the turtle will either decide that everything looks good and begin to dig the nest, or she will decide instead to return to the sea without nesting and try somewhere else later that evening or the following night.

The precise location of the actual nesting site appears to be quite carefully chosen. It must be high enough on the beach to avoid damaging inundation of the developing eggs during subsequent high tides. But if the nest is sited much beyond that point, the developing hatchlings will be exposed to increased risk from other dangers. The longer the journey the new hatchlings will have to make to reach the comparative safety of the sea, the greater the peril. Thus, the precise choice of the nest site is criti-

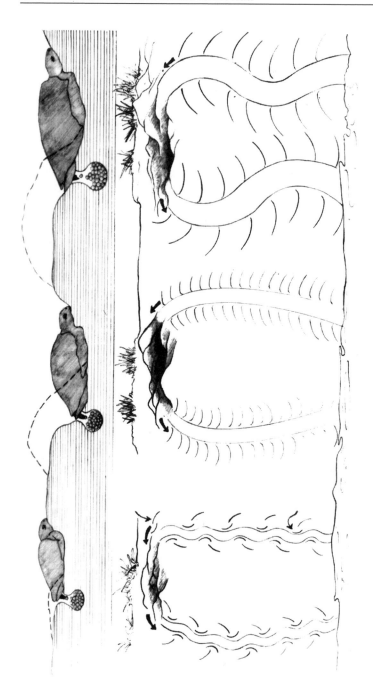

8-4. *The tracks left by loggerhead (left), green (middle), and leatherback (right) sea turtles are distinctive.*

cal, and is not taken lightly by sea turtles. The preferred site of most loggerhead nests in Florida seems to be a short distance from the base of the dune.

Once chosen, the nesting site is first swept clear of debris. The nest proper will consist of two major components; a body pit that provides a lowered platform from which the eggs are eventually deposited, and an egg cavity that will actually house the eggs. The turtle employs a scooping motion of all four flippers combined with rotation of the entire body to excavate the body pit. When completed, the pit takes the form of a shallow depression just large enough to house the turtle (Figure 8-5). It is not very deep, as the top of the carapace generally still extends from the pit above the level of the surrounding sands.

Next, the turtle proceeds to dig an egg cavity by using the rear flippers only. This limits the depth of the chamber to the distance below the body pit that may be reached by the rear flippers, and presumably larger turtles are able to dig somewhat deeper nests. The egg cavity is shaped somewhat like a vase, wider at the base than the top. After the eggs have been deposited into the cavity along with a protective mucous, the cavity and body pit are refilled and carefully packed. Loggerheads may deposit more than one hundred eggs in a single nest. Finally, the turtle conceals the nest using her flippers to redistribute the surrounding sand in a haphazard pattern over the top of the site. The process of refilling and concealing the nest may take as much as twice as long as the process of digging it. Loggerheads usually nest more than once during a single season, with approximately two weeks between successive nestings.

In about 50–60 days the young turtles hatch. The process of hatching and subsequent escape from the egg cavity is a somewhat complex and fascinating phenomenon in its own right. Only rarely are individuals able to reach the surface on their own. The general rule is for the nestmates to work together as a coordinated group to achieve freedom. Once at the surface, they must quickly find their way to the sea. Hatchlings use a very simple behavioral response to find the water; they simply move towards the brightest area of the horizon. Under natural conditions, this

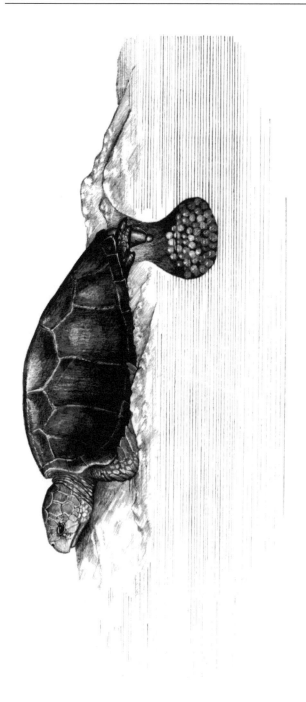

8-5. *The internal structure of the nest of a loggerhead turtle. The egg chamber is annexed to the body pit above.*

will be the sea. When artificial lighting is placed landward of a turtle nesting beach, however, the hatchling's instincts lead them in the wrong direction, often to their deaths.

Upon reaching the sea, the nestmates become separated from one another and are left entirely to their own devices to make their way to deep water. The length of the seaward journey varies along the east coast, as the waters of the Gulf Stream meander farther or closer to shore at different locations. In the area of heaviest nesting (central east coast) the hatchlings may travel about 20 miles or so. Further south, this distance may be substantially less. During this exodus, the young turtles do not stop to feed or rest, for the shallow waters of the continental shelf are fraught with predators and time spent here is a time of great risk. After several days of continuous swimming, surviving hatchlings reach their destination—lines of *Sargassum* seaweed drifting in the warm, clear, deep waters of the Gulf Stream. The journey is perilous, and only some of the original contingent of nestmates successfully negotiate the passage to reach the protective and nourishing environment of drifting weed. Of those that are successful, only a few will survive the perils of the next twenty or more years and reproduce. Surviving females may expect about thirty years of active reproductive life.

Green Turtle

The green turtle (*Chelonia mydas,* Plate 38) occurs throughout much of the tropics, with major populations in the Pacific, Indian, and Atlantic Oceans. Its name is somewhat deceptive, as green turtles actually are darkish brown in color. The species received its English name from early European hunters more interested in the green cartilage, or *calipee,* so prized as a delicacy by Europeans of the last century. Adult females are of about the same general size and weight of the loggerheads—around 3 feet and 300 pounds.

Florida does not host anywhere near the number of nesting green turtles as do its main nesting beaches bordering Venezuela's Aves Islands and along the Atlantic coast of Costa Rica, but still an estimated 400 to 800 females come ashore here each year to deposit their eggs. As with the loggerhead, the beaches of Florida's east central coast are the most popular nesting sites. Between nesting periods, green turtles may journey long and far to favored feeding grounds.

Seagrass is the primary food source of the herbivorous adult green turtles, and this makes them unique because all other species of sea turtles are carnivorous as adults. The most heavily used feeding area in the Caribbean region lies off the Atlantic coast of Nicaragua, where an extensive shallow continental shelf supports vast seagrass meadows. The general nesting behavior of the green turtle is quite similar in most respects to that described for the loggerhead.

Leatherback Turtle

The leatherback (*Dermochelys coriacea*) is the largest of the sea turtles, reaching well over twice the length and weight of loggerheads and greens (Plate 39). This animal is easily recognized by the seven prominent ridges that run the length of its black carapace. Another distinctive characteristic, the unusually long and powerful front flippers, enable the leatherback to traverse thousands of miles of open ocean and dive to great depths. Dives to well over half a mile have been documented, and are believed to be associated with feeding activity.

The leatherback is the widest ranging of the sea turtles, occurring in all major oceans and even venturing into the cold waters bordering arctic and antarctic seas. Leatherbacks feed primarily on jellyfish and other similar large animals of the plankton. Major nesting areas are located in the Indo-Pacific region, French Guiana, and along the Pacific coast Mexico. Fewer than 50 nesting leatherbacks are recorded each year in Florida. The distribution of these seems to indicate a preference for steeply sloping beaches near deep water.

Hawksbill Turtle

The smallish hawksbill turtle (*Eretmochelys imbricata*) possesses a delicately patterned brownish-gold carapace of unusual beauty, a major factor in its demise at the hands of man. Throughout its range, which spans the world's tropical and subtropical seas, it is widely hunted for its shell, which is made into jewelry or decorative trinkets. Hawksbills are quick and agile swimmers, and frequent coral reef areas where they find both food and sheltered resting spots. Adult hawksbills feed mainly on the abundant sponges that grow in shallow warm seas.

Hawksbills are usually observed nesting singly or in small numbers on isolated beaches, many of which are located on small remote islands. In Florida, very few nesting individuals are observed today, although reports indicate that these animals used the state's beaches for this purpose in far greater numbers during the last century. Scuba divers have the best chance of gaining a close look at a hawksbill turtle in Florida, and frequently report observing them swimming about or resting in reef areas, particularly off the southeastern coast and the Keys.

Kemp's Ridley Turtle

Although virtually all sea turtles are officially recognized as either threatened or endangered, the survival of no other appears more precarious than that of Kemp's ridley turtle (*Lepidochelys kempi*). This rarest of the sea turtles has but a single nesting site, located on the Mexican Gulf Coast a few hundred miles south of the Texas border. A population that in 1947 was estimated to contain some 20,000 nesting females has now declined to just several hundred. Conservation efforts have been underway for some time, but have met with little success.

Kemp's ridley is small by sea turtle standards, generally averaging just over two feet in length and weighing under 100 pounds. Its carapace is a light olive color and the plastron yellow. Although they do not nest in the state, Kemp's ridleys are sometimes seen swimming near Florida's shores, perhaps searching for food or migrating between the Gulf of Mexico and northerly feeding grounds off the eastern coast of the United States. Kemp's ridleys favor crabs as a food source, and during the summer months are frequently observed in the Chesapeake Bay feeding on blue crabs.

9

The Keys: Florida's Tropical Isles

The shorelines and seashore life of Florida's Keys are those of tropical Caribbean isles, and differ sufficiently from those of peninsular Florida to warrant a chapter in themselves. The islands and the surrounding seafloor with its overlying waters are all interconnected parts of a tropical ecosystem that shares many essential common elements with those found along island and mainland coastlines throughout the Caribbean region, including Central America. By virtue of their immersion in the tropical waters of the Gulf Stream, as well as their island nature, in many ways the coastal marine environments of the Florida Keys have more in common with those of Cuba and the Bahamas than they do with the shorelines of the much closer U.S. mainland described in earlier chapters.

THE MARINE ENVIRONMENT OF THE KEYS

The marine environment of the Florida Keys, from the Gulf Stream to the island shores, is a world dominated by coral reefs, coralline sand, seagrass meadows and mangroves, all integrated components of a single vast tropical ecosystem. The shore itself is mostly composed of exposed coralline rock or lined by mangrove forest; there are very few sandy beaches, and those that exist are made up of pure coralline sand produced mainly through the bioerosion of skeletal remains of once living organisms.

Although the Keys share a great deal of similarity with Caribbean islands in terms of the general kinds of marine habitats present and their general pattern of distribution (Figure 9-1), there are also some noteworthy differences. First, the Keys are climatically at the northern limit of the tropical western Atlantic, and therefore lack some of the species characteristic of more centrally located coasts. Also, on many Caribbean islands, the entire inshore ecosystem is contained within a few hundred yards of the shore, after which the seafloor plummets to great depths. In the Florida Keys, the shallow inshore environment spans many miles. Most Caribbean islands are bounded on all sides by open seas, whereas the open ocean lies to only one side of the Florida Keys. The other side is composed of a shallow bay-like environment that gradually grades into the continental mainland, a source of great quantities of freshwater and its burden of sediment and pollution. The Keys themselves form somewhat of a protective barrier that greatly reduces the mixing of the turbid waters of Florida Bay with the clear offshore waters brought in by the Gulf Stream. Nevertheless, considerable mixing still occurs and the general water clarity of the Florida Keys, particularly nearshore, is considerably less than that generally found near oceanic Caribbean islands.

These are all factors in determining the nature and distribution of marine life any given area, for many tropical marine species are highly sensitive to temperature as well as a number of aspects of water movement and quality. Thus, it is no surprise that although similar to Caribbean islands in general ecology, the Florida Keys may in some ways be considered quite unique. In recognition of the value of these priceless natural resources, most of the waters surrounding the Florida Keys have been declared a National Marine Sanctuary, and are actively patrolled and managed as such. Sanctuary headquarters offices are maintained in the upper, middle, and lower Keys and offer an excellent source of general information about the ecology and marine life of the area, as well as current regulations regarding uses of these resources.

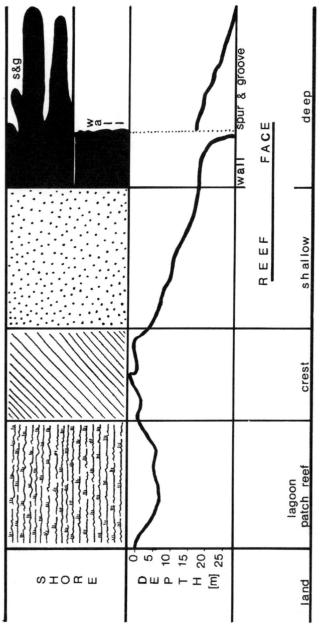

9-1. The typical distribution of Caribbean inshore marine habitats. The reef crest divides the inshore environment into a protected lagoon (shoreward) and the exposed reef face (seaward).

MARINE HABITATS OF THE KEYS

Coral Reefs

Coral reefs are found in tropical areas of all major ocean basins. The variety of life forms inhabiting coral reefs far surpasses anything else the sea has to offer. Hundreds of different fishes, plants, and invertebrate animals may all be found within a single reef area. Huge though they may sometimes be, coral reefs are composed of individual animals that are quite small—few exceed the size of a pencil in diameter. Corals are cnidarians— close relatives of the much larger and generally more familiar sea anemones—and are really quite similar to these in basic structure (Figure 9-2). The most notable differences are that reef-building corals are much smaller, and secrete a hard external skeleton made of limestone extracted from seawater.

Corals only grow in clear shallow tropical seas. A primary reason for this is the coral's need for sunlight, for within the living tissue of the coral animal are found tiny plant cells called *zooxanthellae*. A type of alga, these provide the coral with food, exchanging the products of photosynthesis for nutrients provided by the animal. In truth then, corals are composite organisms, part animal and part plant. Each individual has what amounts to its own internal grocery store, and most of its energy requirements are fulfilled in this way.

Corals also feed on small creatures floating in the sea. These are captured by the tentacles and passed to the gut for digestion, with the resultant nutrients shared by the plant cells. In most species the tentacles are used only at night when they are relatively safe from hungry fish (Plate 40). By day the tentacles are retracted into the safety of the rocky coral skeleton (Plate 41).

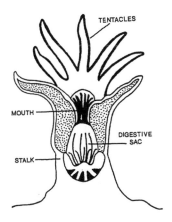

9-2. *A coral colony is composed of the skeletons of countless individual polyps.*

The tiny individual coral animals, or **polyps,** form interconnected groups called colonies, with each succeeding generation building its home upon the foundation left by the last (Plate 42). The colonies in turn are typically attached to colonies of other species, forming larger assemblages, or reefs (Plate 43).

A particular group of plants, the red algae, contains members that, like the corals themselves, also secrete limestone skeletons. These grow amid the coral colonies and serve to cement the coral colonies into a cohesive structure. In that capacity, the red algae in themselves may form a substantial portion of the solid framework of the reef. Upon this rocky base grows an assortment of sponges and other animals and plants, giving the entire structure even greater complexity and variety, and completing the formation of the wondrous entity we call the coral reef.

Each kind of coral forms colonies of distinctive shapes and sizes, and although these are subject to some modification by local conditions, most species may be readily recognized by the characteristic appearance of the colony. Some species, however, are highly similar, and these require close and careful scrutiny for proper identification. In all, just over 50 species of reef-building corals are found in the western Atlantic region.

As well as the reef-building type of coral previously described, the term "coral" is also used to include a group of similar animals that secrete flexible rather than stony skeletons. These are the *octocorals,* and they are a prominent feature of Caribbean reefs and hardgrounds (Figure 9-3; Plates 5 and 26).

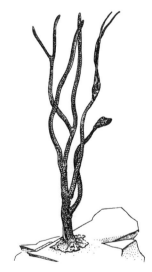

On a larger scale, coral reefs occur in a variety of types. The shallow platform that forms the seaward portion of the marine environment of the Florida Keys is bounded towards the outer margin by an intermittent series of a type of coral

9-3. Gorgonians grow from hard substrates from shallow to deep waters.

reef known as *bank reefs,* because they occur along the border of a marine platform or bank. These reefs roughly parallel the shoreline for nearly the entire length of the Keys at distance ranging from about 4–7 nautical miles from shore. Bank reefs extend virtually from the surface to depths as great as about 120 feet. Beyond the bank reefs, the waters deepen into the Florida Straits, the highway of the Gulf Stream. The line of bank reefs stands like a wall, absorbing and dissipating the energy of incoming waves, and represents a rough dividing line between more protected lagoonal waters to shoreward, and the exposed waters and open oceanic conditions found seaward.

The relatively calm waters shoreward of the protecting system of bank reefs are a rich habitat for some of the most heavily exploited animal life of the region, including spiny lobster, crabs, conch, and fin fish. This is the province of another type of coral reef known as the *patch reef,* isolated coral worlds surrounded by seas of sand and seagrass (Plate 44). Patch reefs may be the size of a small car, or larger than a football field, and may be composed of just a few, or a wide variety of corals. Most are intermediate between those extremes, perhaps averaging 20–70 meters in diameter, with a characteristically round to oval shape. The serene waters above patch reefs are ideal for snorkeling, and by far the best place to become first acquainted with reef life. Larger patch reefs are generally occupied by a rich assortment of fishes.

Small coral colonies may be found growing quite close to the shore in many parts of the Keys, and may be readily observed with nothing more than a dive mask and snorkel in wading depths. Remember that all coral occurring in these waters is protected by Florida and federal law, and must not be touched, damaged, or collected.

Hardgrounds

The habitat called hardgrounds, or live bottom, represents a second major type of reef environment to be found in the Keys. Hardgrounds are low platforms of eroded limestone covered with a living carpet of sponges, octocorals, and encrusting plants and animals. Frequently rising no more than a meter or so from a sandy sea floor, these areas are formed from the exposed bases of ancient coral reefs that have been worn by the forces of nature until little remains. The hard substrate is perforated by numerous cavities—holes and crevices that give shelter to a multitude of animals. The platform also provides fine attachment sites for octocorals and sponges, and the long-dead reef is thereby transformed into a fairy garden, alive with waving sea fans and colorful reef fishes.

Hardgrounds may occur as scattered mounds among the patch reefs of the lagoon, or as large terraces in shallow or deep water. They provide an added dimension of habitat diversity to the inshore environment, and are a great place to explore in their own right.

Seagrass Meadows

Seagrasses are a group of marine plants that form their own distinctive habitat, complete with resident assemblages of other plants and animals (Figure 9-4; Plate 45). Seagrass beds are most often found interspersed with patch reefs in calm protected waters, where they serve as primary foraging grounds for many species of fishes and invertebrates.

Seagrass beds have little physical complexity. There is scarce shelter here above the grass blades, except for the occasional sponge, octocoral, or small coral head. The animals active here by day are mainly small invertebrates and fishes that rely heavily on concealment—either through camouflage or burrowing. Only under the cover of darkness may larger animals venture forth and find a measure of protection from the swift predators that patrol this domain. Many of these animals feed in the open at night and take shelter on the reef by day, where they release the waste products of the digestive process, organic materials rich in nitrogen and other essential nutrients. Such processes thus play a direct role in the transfer of nutrients and energy between seagrass beds and coral reefs, and underscore the pivotal role of interactions between very different kinds of habitats contained within the larger framework of the same overall ecosystem.

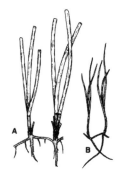

9-4. *Common seagrasses of the Florida Keys:*
(A) Thalassia testudinum;
(B) Halodule wrightii.

Mangrove Forests

Mangroves are a rather unique group of plants that line the shores of many reef areas and contribute to the overall productivity and biotic diversity of reef ecosystems (Plates 1 and 46). One of the rare terrestrial plants able to tolerate direct immersion in seawater, mangroves also play a key role in creating land from the sea by trapping and consolidating sediments that would otherwise be swept away and dissipated. Mangroves are true pioneers, often instrumental in the building of new islands and shorelines.

Mangroves provide both food and shelter for an amazing variety of plant and animal life. The unusual and complex root system forms a nursery ground for young reef fishes, and a substrate to which all manner of algae and invertebrate animals may attach. Falling leaves and nesting birds add nutrients to local seawater, enhancing plant growth. Some of this energy is eventually transferred through a series of intermediate links to creatures of seagrass beds and coral reefs.

Mangroves are found on both the ocean and bay sides of many of the Florida Keys. They are protected by law, and efforts are intensifying to conserve as much as possible of these vital marine plants from further destruction at the hands of coastal development.

Exploring Marine Life in the Keys

Sandy beaches are at a real premium in the Keys. By far, the most extensive and best natural sandy beach is found on Bahia Honda Key, and is managed under the protection of the Bahia Honda State Park. While traditional "high and dry" beachcombing methods will allow you to explore some of the seashore life to be found along the beaches, coralline rocks, and mangroves that line the shores of the Florida Keys, to see the real treasure of these islands one needs to look below the surface of the surrounding seas. Put on a mask and snorkel and explore the shallow waters near the shore at your own leisure. For exploring further out, Scuba and snorkeling trips, as well as glass-bottom boat excursions, are available at most reasonable rates on a daily basis throughout the Keys. A visit to this part of the state is not complete without taking such an opportunity to view the world of the living coral reef first-hand.

Bibliography

Alevizon, W., *Caribbean Reef Ecology,* Pisces Books, Houston. 1993. (An introduction to Florida's coral reefs.)

Andrews, J., *A Field Guide to Shells of the Florida Coast.* Gulf Publishing Co., Houston. 1981. (Identifies and describes almost 300 mollusks of the Florida coast.)

Carson, R., *The Edge of the Sea,* Houghton Mifflin Co., Boston. 1979. (A beautifully written account of the natural history of the seashore by a prize-winning author.)

Chaplin, C. G., *Fishwatcher's Guide to West Atlantic Coral Reefs,* Harrowwood Books, Newtown Square, PA. 1972. (An excellent identification guide for snorkelers of south Florida and the Keys. Available on waterproof paper.)

Fotheringham, N. and Brunenmeister, S., *Beachcomber's Guide to Gulf Coast Marine Life,* Second Edition, Gulf Publishing Co., Houston. 1989. (Particularly valuable identification guide to seashore "critters" of Florida's northwest coast.)

Fox, W. T., *At the Sea's Edge,* Prentice Hall Press, New York. 1983. (A clearly written and beautifully illustrated general introduction to coastal oceanography.)

Kaplan, E. H., *A Field Guide to Southeastern and Caribbean Seashores,* Houghton Mifflin Co., Boston. 1988. (A useful general field identification manual.)

Morris, P. A., *A Field Guide to Shells of the Atlantic and Gulf Coasts and the West Indies,* Houghton Mifflin Co., Boston. 1973. (For those particularly interested in shelling.)

Index

Boldface page numbers indicate figures/illustrations. Scientific names are in *italics*.